CR
D0461890
OVER

CROSSING OVER

One Woman's Escape from Amish Life

Ruth Irene Garrett
with **Rick Farrant**

HarperSanFrancisco
A Division of HarperCollins*Publishers*

HarperCollins books may be purchased for educational, business, or sales promotional use. For information please write: Special Markets Department, HarperCollins Publishers, Inc., 10 East 53rd Street, New York, NY 10022.
HarperCollins Web site: http://www.harpercollins.com

HarperCollins®, ®, and HarperSanFrancisco™ are
trademarks of HarperCollins Publishers, Inc.

FIRST EDITION

Library of Congress Cataloging-in-Publication Data has been ordered.
ISBN 0-06052-992-X (pbk.)

03 04 05 06 07 RRD(H) 10 9 8 7 6 5 4 3

To my mother in heaven, who never lost the faith

—Rick Farrant

To God, all honor and glory; to my mother, the bond of love we share I will carry in my heart forever

—Ruth Irene Garrett

Foreword

One Lord, one faith, one baptism, one God and Father of all, who is above all, and through all, and in you all. But unto every one of us is given grace according to the measure of the gift of Christ . . . Let all bitterness, and wrath, and anger, and clamour, and evil speaking, be put away from you, with all malice: And be ye kind one to another, tenderhearted, forgiving one another, even as God for Christ's sake hath forgiven you.

—EPHESIANS 4:5–7, 31, 32

The Old Order Amish have many traditions in their Ordnung (church rules), among them no picture taking and no possessing photographs of people or family members.

Therefore, I have very few photographs or family albums with which to reflect on my past. I must rely on recollections and memories.

When a person remembers or reflects on his or her past, it should bring back beautiful and wonderful moments in time, as well as childhood dreams. While I have many pleasant memories of my time as a member of the Amish religion, I also have unpleasant and sad memories that I wish not to remember but cannot forget.

I understand what is needed to deal with the truth and the past, that being love. The love I share with my husband and my love for my Amish family have enabled me to be strong and tell my entire story as honestly and truthfully as possible. My cowriter, Rick Farrant, was there, too. He was professionally persistent and sometimes asked difficult questions, but he was nevertheless always aware of my desire to protect my family. The family I left. But also the family I love.

There are more than 235 Old Order Amish communities in twenty-three states and parts of Canada—about 150,000 people in 1,100 church districts. The Old Order is the strictest of the Amish groups and it, along with the more liberal Amish and Mennonites, arose from the sixteenth century Western Europe Anabaptist movement, influenced in large part by two former Roman Catholic priests turned reformers—Martin Luther and Menno Simons. Both challenged the power of the pope and Catholic Church doctrine.

Jacob Ammann, a Swiss Mennonite bishop, founded the Amish in 1693 when he became dissatisfied with the Mennonite Church, believing it did not carry shunning far enough and did not adequately separate itself from the rest of the world. Later, Ammann excommunicated himself from the Amish Church in a failed attempt to rejoin the Mennonites.

I come from an Old Order Amish group in Iowa, and our dress, our buggies, our way of life were quite different from those of another Amish group less than a hundred miles to the north. My story, then, is rooted in the ways of one particular Amish community and does not necessarily reflect the ways of all Amish.

That said, there is one unspoken rule among all Amish groups that for centuries has kept the Amish a mystery to the outside world. They avoid talking about their faith and traditions to the English, otherwise known as all people who are not Amish. They do not do taped or television interviews. They hide their faces from cameras and do not become too friendly with outsiders, lest they should be tempted to leave the Amish and become English.

This may sound absurd, but it is not when you have been told all your life that very few English can be trusted. At the same time, by limiting contact with the English, the Amish avoid having to answer any questions pertaining to what goes on behind closed doors.

If one leaves the Amish and talks about the inner workings of the Amish or the hardships he or she experienced while being Amish, the Amish will say that person is trying to retaliate and is blaming the Amish for his or her shortcomings.

I am doing neither; rather, I am trying to foster a better understanding among all Christians and non-Christians, that we may learn how to unite under God's mercy.

Leaving the Amish is perhaps the most serious offense one can commit. From the time you are born until the day you die, you are reminded by the leaders of the church that you are privileged to be Amish, that the world outside the Amish is evil and corrupt. One grows up thinking the Amish are the only ones with a real chance of going to heaven.

As long as I stayed in the church, worked hard, did my daily prayers, was humble, and always followed the Ordnung, I might have a chance of going to heaven. But leaving the Amish meant I was no longer of the church, I was in the ban, and I was cut off from God. To them, if I die this way, I have doomed my soul to hell and have no chance of salvation. Being in the ban also means one is shunned and can have little or no contact with other Amish.

With these threats of damnation, it is not hard to imagine why Old Order Amish live in fear and intimidation. It is not hard to understand why they do not talk freely.

I am not sure my family will ever understand why I decided to tell my story. It is not meant to hurt them. It is told with the belief that God loves all of his children and wants all of them to be saved, not just a few. Any denomination that believes otherwise has already missed the point.

I look at my father and I see a man who is unyielding in his convictions and stern in his punishment. I see a man who is a product of his father's beliefs and Amish traditional teachings. My mother, on the other hand, is as kind and loving as my father is strict. My brothers are good, strong men whom I love. For my sister, I wish I could tell her that she is loved and will always be loved by me.

I love my mother and father, and I appreciate everything they have done for me. They have taught me many good values. Still, I so wish they would try to understand that I wish to be where I am.

I have not forsaken God; rather, I have embraced my beliefs in God and Jesus more than ever. The church I now attend is an extension of my faith in God and I have taken my faith far beyond anything the Amish would ever dare to do. I did this with the help of my husband, Ottie, Rev. James Bettermann, and the entire congregation of the Holy Trinity Lutheran Church. I have formed a new foundation for my faith in God and his plan of salvation by grace—a new understanding, a new beginning, a new life through truth, patience, and love.

This book tells about my experiences growing up Amish, my husband, my hopes, and my dreams. I feel compelled to tell my story, hoping it will help anyone with similar struggles of their own. If I have hurt anyone in so doing, I ask for their forgiveness and prayers.

—Ruth Irene Garrett

One

February 5, 2000

I had barely made it inside the door of the old, two-story white farmhouse when my mother fell into me, put her head on my shoulder, and began sobbing. "Everything's all right," I whispered, wrapping my arms tightly around her. "It's all right."

We hadn't seen each other in more than three years.

I had returned to my childhood home in Kalona, Iowa, to see about my mother's health. I'd heard she had open sores on her legs, and a quack doctor had ordered a bizarre diet as a cure: chicken, fish, wheat biscuits, popcorn made with canola oil, no sugar, no fruits. I also was intent on proving to her—and to my father—that I was okay after leaving home, going into the world, and becoming a bride.

We'd exchanged dozens of letters in which she and the rest of my family had pleaded for my return and I had promised I was in no danger. But none of my words had made a difference. They were convinced I had abandoned God and sold my soul to the devil—that I was under some kind of spell that had prevented me from returning. They had implored in words blind with grief and dramatically scripted to touch my insecurities. And always there was the desire that I forgive them for anything they might have done to encourage me to leave. "How empty your place in the family is!" my mother wrote in a letter two years after I'd left. "I'm alone in the house so I can let the tears roll and not offend. It would be much lighter and easier to get ready for this family gathering if you would be here to help like you used to be.

> "Oh Irene! It gets so lonesome without you and it's hard to go on; the waiting is so painful. I try to go on for the rest of the family's sake. I'm very, very sorry that something like this happened and hope I can be forgiven. I do not have much in earthly possessions, so our children are very dear to me and that is why it hurts so deeply. . . . May the Lord grant you wisdom to see the truth in all matters."

They often invoked the name of God. In a manipulative way, it seemed.

The Bible doesn't say anything about conditional love or guidance. It says that if people follow God, they will have his blessing. And no matter what I might have done, I had not turned my back on him, as my parents would frequently suggest. If anything, I had more fully embraced his comfort, forgiveness, and strength. That is what was allowing me to love my family still.

My heart raced, my face flushed, and my hands turned cold as I tried to calm my mother. But I remained strong, even though it had been a difficult twenty-four hours leading to the reunion. There had been the persistent headaches, the bouts of nausea.

The day before, I'd left Kentucky on my first commercial

plane flight ever and flown to Chicago, where I was met by an aunt and uncle who'd driven from Indiana. Together, we had made the six-hour drive to Kalona, eighteen miles southwest of Iowa City. We checked into the forty-nine-dollar-a-night Pull'r Inn Motel, two miles from the Gable Avenue farm where I'd grown up. I saw the eighty-acre property amidst the gentle swells of the eastern Iowa landscape as we traveled into town along pink-topped Highway 1. Everything was as it had been. The gray and silver silo, the silver windmill, the long, sloping barn roof, the white martin house on the pole.

My biggest concern was that people would discover I was in town before I got a chance to see my family. That would have been disastrous. My father would have summoned a gaggle of ministers, and they would have confronted me, surrounded me with feelings of guilt, and hounded me into submission. Because that's what they do to fallen members of their flock. They badger them. Intimidate them. Shame them. And demand obedience. I would have been mortified.

Signing into the nondescript motel at the juncture of Highways 1 and 22 provided the first challenge to protecting my identity. I wondered if I should use my real name or an alias. Then, either from exhaustion or a commitment to honesty, I wrote the name I had assumed after leaving home. Irene Garrett. My real name. My married name.

Later that evening, I met up with a woman friend who helped shield me. When the former fire chief looked inquisitively at me as we dined in a pizza restaurant, my friend made a point of making no introduction. She didn't spill the secret to our waitress, either.

By the time I got back to the motel, I was still undetected. But I was more than frazzled, and I didn't get to sleep until after midnight—after doing a good deal of praying and writing in a notebook I had brought along: "It's almost like I've entered communist territory, where I have to sneak in and out before too many people know I'm there. Before they have all . . . the authorities gang up on me. Almost like a fugitive. . . . Going to visit your parents should be easy and something you anticipate, but it's the hardest thing I've done in my life." I was twenty-six years old.

The morning dawned quickly—and not without incident. I dressed conservatively out of respect for my parents—including wearing a small head covering—joined up with my aunt and uncle, and we went to the Kalonial Townhouse Restaurant for breakfast, where one can order a local favorite of liverwurst, fried mush, and coffee for $4.75 or an Iowa chop for $7.25. No sooner did we arrive at the popular gathering place than a group of people I recognized walked in. One of them, I could tell, was a former neighbor. I could see them peeking at me and hear them talking among themselves, and one said, "Well, I think that's one of Alvin T's daughters."

Moments later, one of them came by our table and asked, "Aren't you teaching school?"

"No," I said, "that's my sister you're talking about."

"Well, are you the one . . ." I finished the sentence for him: ". . . that ran off to Kentucky? Yeah, that's me."

He laughed and we chatted, but now the secret was out. And it wasn't long before a waitress, who'd been inspecting me from the chrome-hooded buffet, blurted: "Oh, Irene, is that you?" In orderly fashion, three more waitresses, all of whom had once been acquaintances of mine, eventually made their way to our table. One of them asked if my parents knew I was in town. "No," I said. But soon they would know. In about half an hour.

We got to the farm at 9:00 A.M. I had spent the drive fighting the urge to cry or tighten up. Being in control was so important to me I'd almost lost it. Somehow, I managed to keep my wits. The farm looked deserted as we pulled in, much like it had looked two and a half years earlier when I'd tried to visit. That time, the family had gone fishing for the day. This time, I braced myself for another disappointment.

My aunt went to the front door and knocked and, as we stood nervously next to the green, wooden front-porch swing, I heard a chair leg scrape the floor inside. Someone was home. My mother.

It took me ten minutes to get her calmed to the point where I could lead her to the beige sofa in the living room. Then I held her hand and waited for the questions I knew would come. Questions that were so real to her, and so ludicrous to me.

She began without warm-up: "Is he treating you all right?"

"It couldn't be better," I replied.

"Well, I just hear all these bad things," she said, looking thinner than when I'd last seen her. "These horrible things. Just last Thursday, I heard another rumor."

"What was that?" I asked.

"That he's got guns mounted in the house. That he can shoot you if you try to leave him."

"All this stuff you hear is not true," I said. "It's not true."

But there was more. She'd heard that I once had a bracelet on my arm that allowed my husband to track me, and that I still had a microchip in my wrist that branded me with the mark of the beast—666.

"What about this microchip?" she asked. "Do you have one of those?"

"No," I said.

"Well, don't you ever get one of those, because you know what the Bible says about that."

It is never easy leaving the Amish. Especially if you've left to marry an outsider.

Two

It seems so hard to write when I feel so broken up. It's so hard to take that you have left. It's hard to realize that even the sun could rise again . . . Come! Come! Before it's too late . . . God can heal a broken heart, but he needs all the pieces. So please come!

—Letter from Martha Miller (Mom)

I guess you could say I had a normal childhood, considering I didn't know any other. I was the fifth of seven children, born Ruth Irene Miller on January 31, 1974, roughly eight months before Richard Nixon resigned. The oldest child was Elson. Then there were Bertha, Wilbur, Benedict, me, Aaron, and Earl.

Actually, there had been two others. A brother, Tobias, died at age three in the basement of our home in an accidental fire. A sister, Miriam, was born with a deformed skull and died several hours later.

They are buried side by side in Peter Miller Cemetery, a barren rise of shin-high gravestones where the dead are serenaded by meadow larks and protected by a crooked, rusting metal gate.

My brother's stone reads: "Tobias A. Miller, son of Alvin & Martha T., 1964–1967, Age 3 yrs., 29 da." My sister's: "Miriam

A. Miller, infant dau. of Alvin T. and Martha T., B. Sept. 3, 1980, lived 3 hrs."

Tobias's accident happened before I was born. He had poured gas into a spray can customarily used for fly repellent. He got too close to the water heater's pilot light and the gas exploded. The heat from the fire was so intense it blistered paint on the basement's walls.

My father would later say it was a blessing that Tobias died. Satan, he said, would never again tempt him to do wrong.

Miriam, a doctor said, may have been the victim of exhaust fumes coming from a leaky conduit that led from the washing machine to a gas engine. The pipe had six cracks in it, and Mom would frequently get sick while doing the wash.

Miriam's body was put in a little white Styrofoam box that was placed on a dresser in my parents' bedroom. It stayed there for three days while we waited for relatives to arrive in Kalona for graveside services.

Benedict and I once went into the bedroom and tried to open the lid of the box, but it had been sealed shut. At the time, I don't think we fully understood what was inside.

Of course, there were a lot of things we didn't understand then; couldn't possibly have understood then.

As Old Order Amish children, we were taught that we were the privileged ones, chosen by God to do his work and the only ones who stood a chance of being saved. We therefore were forbidden from doing missionary work outside the community.

We were also warned that everything outside our world—otherwise known as English—was evil, inhabited by thieves and liars.

Amish children—along with Amish wives—were, and still are, a subservient class. Children are valued more for their work habits than their developing personalities, and any money they make before they turn twenty-one must go to their parents. Wives exist to care for the children and to serve their husbands. Nothing more.

That's not to say we didn't have fun. In our spare time, two of my brothers and I would run barefoot in the van Gogh fields behind our farm and hunt butterflies for my collection. We didn't have a net, so we'd sneak up on them and, just when their wings shut, we'd slip our fingers over them. We put them to rest by dipping their heads in tiny vials of gasoline.

I was attracted to the beauty of the butterflies. And while I felt sorry that they had to die, I figured that meant I could enjoy them that much longer. Such is the way of life and death—on a farm.

Sometimes we'd also go hunting for bird eggs to destroy, targeting the nests of blackbirds and cowbirds. They were considered nuisances because they robbed the nests of other birds.

It was a fascinating exercise. You'd have to look up and watch the behavior of the birds. When they circled above your head, scolding and carrying on, you knew you were close. When they backed off, you knew the trail was cold. Sometimes they'd get really close to our heads, but we never paid much attention.

Again, it wasn't that I didn't like birds; I'd spend hours listening to their whistles and songs so I could identify them by sound. It's just that on an Amish farm, animals are viewed more as commodities or pests than pets or natural wonders.

One time, an English woman came to our farm and began fussing over a piglet in the hog shed. She kissed it all over despite its foul smell, and I thought, "This lady's lost it. She thinks this stinking pig is a pet or something."

One of our favorite childhood pursuits had nothing to do with animals. We made mud cookies. We'd take water and mix it with dust until we had the right consistency. Then we'd shape the mud into patties, put them on sheets of tin, and set them in the sun to dry. Later, we'd top the patties with seeds from lamb's-quarters. They looked just like sprinkles on a cookie.

We didn't eat them, of course. We'd pitch 'em when we were done.

I also remember riding ponies, ice skating on our pond, playing church and school, competing in a pool-like game called carom, frolicking in the corn crib, hayloft, and silo, and engaging in a bit of role-playing unique to the Amish.

Instead of playing with cars and trucks, as many English boys and girls do, we'd play horse and buggy. My mother would make harnesses and reins out of scraps of denim and we'd take turns playing the horses and drivers, pulling tiny wagons or carts around the yard and reprimanding the horses when they grew unruly.

These sorts of romps had to be worked into busy schedules that included chores around the farm, schooling, and lots of prayer.

Because we were Old Order Amish—the most conservative of the so-called plain people—we adhered religiously to not owning cars, electricity, or phones, considering them too worldly. That meant most of our work was done by hand or with the aid of gas-powered engines.

On school days, my father would awaken us at 5:30 A.M. with a shout up the stairs. My jobs were to help milk the dozen or so cows, help Mom with breakfast, and wash dishes before heading for school about 8:30 A.M. The milk was a cash crop; a man would come by twice a week to collect it.

Breakfasts were usually pancakes with syrup or with sausage or hamburger gravy. Sometimes we'd have hot and cold cereals—oatmeal, cracked wheat, corn flakes, and the like.

About 4:20 P.M., I'd get home from school and gather eggs (another cash crop) from the chicken house, fill the house's lanterns and lamps with kerosene or gas, and help milk the cows again. Winters, I'd gather wood for the stove.

Because lunch was the biggest meal of the day—meat loaf, fried hamburgers, or chicken, mashed potatoes, and vegetables—dinner was light. Maybe soup. Cheese. Homemade baloney. Some fruit and cake.

Before breakfast and at bedtime, we'd kneel on the floor and Dad would lead us in a German prayer. We'd also bow our heads at the table and have silent prayers at the start and finish of every meal.

Then we'd pray at school and in church, which we attended every other Sunday.

The silent praying was baffling to outsiders who'd occasionally visit. A few of them were in the middle of talking when, without warning, our heads would suddenly drop. I can only imagine what they must have initially thought—and felt once they realized what we were doing.

All this praying, the Amish believe, gives them a chance at going to heaven. And it is only that—a chance. Unlike other religious faiths, which virtually guarantee repentant sinners a place aloft, the Amish believe that by works and deeds they might find themselves worthy in God's good graces to go to heaven.

Ask an Amish person if they're going to heaven, and they'll say, "That's not my choice. That's God's choice."

They can never be sure they're going, because they might misstep between now and the moment they die.

It doesn't make sense. Romans 10:13 says: "For whosoever shall call upon the name of the Lord shall be saved." That leaves little doubt.

But I, especially, had to strictly follow the Amish's inflexible beliefs and stern rules of the Amish church (the Ordnung), which govern everything from dress and modes of transportation to dating etiquette and reading habits. My father, grandfather, and two uncles were ministers. Another uncle was a bishop. As such, they were important, respected leaders of the community, and their families, by association, were expected to set high examples.

Amish neighborhoods are divided into districts, and each district has two ministers, a deacon, and a bishop. The ministers do the preaching at Sunday services, usually held in church members' homes. Deacons make sure members don't violate the rules of the community; if they do, the wayward must confess openly in church. Bishops watch over the ministers and deacons, and also handle marriages, funerals, and baptismals.

The Amish believe that if a minister—in this case my father, Alvin T. Miller—can't keep his family in line, he's not capable of keeping the church in line.

Beyond that, there's an inherent code among the Amish that requires people to monitor the activities of others—and point fingers when someone has gone astray. Neighbors tell on neighbors. Children tell on children. No one is safe from judgment.

Among the things closely watched is the attire of women. They must wear a well-defined set of modest clothing, including starched organdy head coverings with eight pleats on either side. Not five pleats or six.

Tradition is the main reason for this edict. The other: Fewer pleats on either side would make the head covering too small.

Women wear the head coverings (scarves are allowed while working outdoors) all the time, including while they sleep, should they feel moved to pray in the middle of the night. Head coverings worn in church are white if they're married, black if they're single. And they must always keep their hair tied in a bun.

They wear lace-up, black leather shoes; thick, knee-high nylon leggings; and long, homemade, double-knit, dark-colored dresses held together with straight pins. The dresses are to be no more than eight inches from the floor.

When they are away from their homes, they often wear a knit cape that's fastened with pins on the front and back.

There are dress codes for men, too, although they have more liberties. They don't have to contend with pins, and they can wear lighter colors.

Men dress in homemade, buttoned, cotton shirts that are often pocketless; zipperless denim pants that button in flaps across the front; suspenders; wide-brimmed straw or felt hats; and black or brown boots.

They are not permitted to layer their hair or grow mustaches. Those who are married have beards. Those who are single are clean-shaven.

The Amish take great pride in such restrained conformity. Their clothes represent their humility, their lack of vanity. And for some Amish, the extreme is most preferable.

There are those—not my family, thank goodness—who believe the dirtier you are and the more worn your clothing, the more Amish you are. They are the riper ones among the Amish.

This attention to plainness is also mirrored in a lack of expression—literally.

The Amish rarely smile or laugh. They frown on excessive hilarity, believing that if something is extraordinarily funny, it must be bad. They are more content taking their religious, agrarian life seriously, living by the motto that the harder it is on earth, the sweeter it will be in heaven.

The Ordnung, a closely protected set of rules that varies among the Amish settlements, ensures such somber conformity beyond clothes and expression. In our settlement, it also forbade Amish to fly in airplanes unless there was a medical emergency, mandated that the inside of buggies be kept plain, prohibited long curtains in windows, and asked that members not have photographs of people in their possession.

It all sounds rather stifling for a child—or anyone for that matter. But if you have nothing to compare it to, it is your life. And later, when you enter adulthood, you still remember things that at least bring a modest smile to your face.

Like the butterflies, the mud pies, the birds—and the swans.

Not long before I left the farm, my mother bought a pair of big, white, beautiful swans for our pond. She got them at an exotic animal sale in town where the proprietors also offered rheas and llamas, guinea pigs, and mice. My father said the swans were too expensive—I don't remember how much Mom paid—but we kept them anyway.

I was ecstatic. I had fallen in love with swans at age thirteen when I saw their images on a greeting card. They were floating on a willow-bordered stream, and they seemed so proud and graceful and serene.

Swans and butterflies, I realized, have a lot in common. They start out as drab cygnets and caterpillars, then blossom into stunningly beautiful creatures. In this way, I was different from some Amish: I held an awe and appreciation of nature.

And I guess I liked the idea that things so plain could turn into things so pretty.

Three

(Wer auf dem fleisch säet wird auch ernten die ewig feuer, schrecklich wo du stehst! So komm zurück und lasz Jesu dich reinichen!)
Who soweth in the flesh will also reap the eternal fire, frightful where you stand! So come back and let Jesus cleanse you!

—Letter from Alvin Miller (Dad)

My first inkling that the Amish way had flaws was twofold. They were subconscious childhood realizations and certainly not cemented until much later. But they were the beginning of my questioning.

The first language Amish children learn is Pennsylvania Dutch. So, if your parents don't want you to understand what they're saying, they speak English. Hence, children are intent on learning English as quickly as possible. But the language puzzle doesn't end there. All of the church services and all of the prayers are conducted in German. I'm not sure the Amish know why it's done that way, other than it's tradition. But I have a hunch it may be to make it difficult for people to closely interpret the Bible. The Amish, who came to America seeking religious freedom, are fiercely afraid of change and therefore not interested in having an enlightened community. Church leaders tell the flock that if they read the Bible too thoroughly, they'll learn more than they should know. The common refrain is: "You'll get too smart for your own good."

It's like the Catholic days of yore when the priest was the only one who read the Bible; people would merely listen to what he told them.

I know of at least two bishops who were excommunicated—"put out," "put in the ban," or "shunned"—because leaders felt they were reading the Bible too much and not focusing on teaching the Amish way.

Some Amish even believe that if someone begins reading the Bible regularly, it's a sure sign they will be leaving soon—that they will finally know the truth and stop listening to Amish sermons.

This austere commitment to German prayer and preaching is so tightly followed that prayers are never spoken from the heart, never personalized. All prayers are read verbatim from a German prayer book. There is no room for multiple meanings to a passage.

The fact that Amish children are formally educated only through the eighth grade is further evidence of the need to keep the Amish people in the dark. Even the subjects in school avoid any discussion of modern technological advances; the Amish prefer instead to focus on basic reading, writing, vocabulary, spelling, arithmetic, English, and social studies, all taught from outdated textbooks.

But it was the language issue that first raised my suspicions about the secretive, rigid society in which I was raised. Even today, my parents write portions of their letters to me in

Dutch or German when they don't want my husband to understand what they're saying.

As a child, I was forever going to the dictionary to see what words meant. I was determined not to be a third party to anything.

Another troubling aspect of the language maze is that there is no word for "love" in Pennsylvania Dutch. It appears mostly in the German *(liebe)* they speak, and only then when they are referring to God.

On the rare occasions when they use the English "love," it is often so wrapped in religious rhetoric that it loses its effect.

"Greetings of love and best wishes in Jesus' name . . ." is the way some written correspondence began.

The other concern I had growing up was the dominant—sometimes cruel—behavior of my father, a stern, unforgiving man with stunning white shocks on his scalp and chin and a cold look of disdain that can freeze a person standing. For as far back as I can remember, my mother, a gentle woman with a round face that flushes easily, was the subject of his persistent ridicule—warranted or not. The children, meanwhile, were held to sometimes impossible standards, and any weaknesses they might have—physical or mental—became the subject of my father's scorn.

In my mother's case, his less-than-considerate attitude toward her may have stemmed from the fact that she had come to Kalona from another Amish community in Kokomo, Indiana. The Amish are often leery of Amish who come from other settlements, and in some cases prohibit couples from different communities from marrying. It did not help that the Kokomo community was considered more liberal than ours.

My father may have felt pressure to treat my mother as an outsider. Certainly, some people in our community felt she didn't belong.

My mother always seemed to be under watch in Kalona; if she didn't dress precisely to form, some Amish would perceive she'd done it intentionally—to snub the community. At home, my father would find fault with virtually everything she did; even with things she didn't do.

He would frequently scold her, and she would often accept the punishment quietly and move on. Sometimes, she would go outside to cry, but we'd never see it. We'd see only her reddened eyes when she returned. I'm sure she felt very alone.

My parents would never talk to us about their differences, and we would never confront them. That simply isn't done among the Amish. Parents don't share their private lives with their children. Children are taught—above all else except God—to "honor thy mother and father."

And honor them we did.

But the constant berating of my mother nevertheless dug scars in me that I still carry today. Mental wounds that fester.

Occasionally, my mother would stand up to my father, and that would be a source of great pleasure to me. And to my mother, I imagine, although she would never say so.

One time, my father came home from selling pigs and announced they had weighed less than they should have, that he'd received less money than he'd expected. He was furious and all but said it was Mom's fault.

Mom, bless her heart, told him: "I didn't feed those pigs. I didn't do anything with them. So I don't know why it's my fault."

My father, still simmering, went outside and slammed the door behind him. I can't be sure, but I suspect he knew how ridiculous he had sounded.

Some of the children fared less well under my father's iron rule. In most cases, it wasn't so much the spankings from his hand or the whippings he carried out with a leather strap. It was the cutting words that accompanied his explosive, unpredictable temper.

"I hope this will teach you to work next time!" he said once while disciplining one of my brothers for not being attentive to chores.

It was not an isolated occurrence. We were all deathly afraid of him, unsure, for the most part, when something was going to send him into a rage. My sister and some of my brothers became so submissive, they walked around with their shoulders slumped and heads down, unable to maintain eye contact with people.

Looking back—and this is hard to admit—I realize I must have behaved the same way at times.

One of the things my father was most insistent about was conserving water, which came from two sources—rain for bathing and a 300-foot-deep, windmill-powered well for drinking.

The rainwater would collect in gutters around the house and funnel to a cistern. A gas-powered motor would pump the water out of the cistern and through the faucets.

My father was chronically worried that if we used too much rainwater, we'd have to borrow from the well supply—and he didn't want to do that. So, we were left taking bucket and basin baths every day—washing our feet, hands, and faces—and full-body baths once a week in the winter, occasionally more often in the warmer months.

By English standards, the body baths were hardly robust. The children rarely put more than three inches of water in the tub, and to this day I still fill it only halfway.

One of my brothers, Wilbur, became so concerned about doing the right thing in my father's eyes that he developed an obsessive-compulsive disorder about turning off faucets, making sure he was clean, and getting every ounce of milk out of the cows, among other things.

It'd take him half a dozen tries before he was certain a faucet was shut, he'd rinse over and over again in the tub, and he'd pull a cow's teats long after the animal had run dry.

It became a vicious cycle that only made matters worse. The more he'd repeat himself, the more people would watch and the madder my father would get. The more my father punished him for his actions, the more he'd do it.

Watching how my father treated my mother—and his children—gave me a jaded sense of family and marriage. Especially the latter.

The Amish believe—implore, really, with binding mandate—that their people should stay married for life. And there is a stigma attached to those women who never marry. They are considered old maids.

Some Amish women would rather get married and be unhappy than be old maids. But I decided early on that, given such a choice, I never wanted to get married. I never wanted to walk the valley that was my mother's and father's relationship.

Four

Last Sunday evening, as I was walking home from the singing, I just simply couldn't keep from crying. Oh, the empty spot you have left behind you! It seems I can't start a song in the singing because I am afraid I can't finish it.

—LETTER FROM BENEDICT MILLER (BROTHER)

The commercial, widely accepted, modern-world notion of femininity—the essence of womanhood and sexuality—is nowhere to be found in the Amish culture.

The dour dress has something to do with that, of course. There isn't a spot of bare skin to be seen, save for the hands and face, and the bare feet in summer. Makeup is forbidden. Amish women shave neither their armpits nor their legs. There are no fancy hairdos, no elegant shoes with high heels, no glittering jewelry hanging from the neck and wrists.

There is also little recognition that boys and girls might be attracted to one another. To the contrary, they are encouraged not to mix and are, in fact, segregated. They sit separately in church—just like the men and women—and do not often spend time together outside the church or home.

Even when the children get older, their commingling is so regulated, it's a wonder any of them develop intimate relationships.

They are not permitted to hug, kiss, or hold hands. To this day, I've never seen my parents hug or kiss.

Dating, which is permitted only after children turn eighteen and have previously joined what is called a young folks group, usually occurs after Sunday night singing. A boy will take a girl to her home in his buggy, and she will choose a snack. Popcorn and juice, perhaps.

The two of them will talk and eat for several hours, always by candle or lamplight, for they are prohibited from spending time with each other in the dark. They won't touch and they won't discuss carnal matters.

Truth be told, they probably wouldn't know what to do anyway. Sex education is virtually nonexistent. There are no discussions with parents about how it's done—or what it might lead to. If a child musters the courage to ask about sex, he or she will likely be told: "Wait till you grow up." Which in an Amish parent's mind means something approaching never.

Approved leisure-time reading materials also offer no clues. Popular Amish magazines like *Young Companion* and *Family Life* carry frothy, harmless stories about the seasons, building homes, settlement histories, swimming, obedience, teaching, and the like. They're also real keen on recipes and poems.

Some settlements have wild periods in which teenagers go out into the English world and sow their oats—the broad path—before being expected to return to the narrow way of the Amish church. Such was not the case in Kalona, however.

My mother did hand me a pamphlet when I was about thirteen that informed me in very clinical terms about the female anatomy, particularly the onset of periods. But there was nothing dealing with the passionate elements of relationships.

We didn't even flirt. If you flirted, people would say you were boy crazy, and being boy crazy was a disgrace, especially among poor or middle-class Amish. It suggested a young woman was desperate to get married.

People would visit from other Amish communities and, upon seeing the boys and girls in Kalona, remark: "Man, they look like they're gonna bite each other."

I can honestly say I never looked in the mirror and wondered if I'd be attractive to a boy. For the longest time, it never even dawned on me that an intimate relationship with someone was possible.

My mind and body were simply not equipped with the same social conditioning—and that must be what it is that compels young people in the English world to experiment sexually. I didn't even explore my own body, believing instead that it should be hidden—even from me.

Some Amish women reinforced this notion by sewing bras so they would diminish, rather than enhance, curves. No points, no uplifts, no extra padding.

My father also influenced me. When he'd see a well-endowed woman, he'd remark, "Man, she's as big as a cow," which I took to mean that big-breastedness was something to be embarrassed about. It wasn't until later that I realized—guessed, anyway—that my father was probably attracted to busty women but was trying to hide his interest.

Everything else aside, I had one strike against me from the start in the femininity department. I was a tomboy, and I was competitive, and that usually meant going head to head with the boys.

In softball, I relished hitting the ball over the fence—just like the boys—or at least hitting it far enough that they'd have to run after it. Just seeing them back up in the outfield when I came to bat was a thrill.

I was just as determined playing carom, and I even won a school tournament with my cousin.

Carom is played on a square board that has pockets, much like a pool table. There are twelve plastic green checkers and twelve plastic red checkers that look a lot like donuts. Four players, two to a team, use white shooters—cue balls—and sticks to put the checkers in the pockets. First side in wins.

I was equally adept at spelling and won several German and English spelling bees.

I also loved to ride horses and ponies—and we're not talking gentle gaits. I'd take them out on the farm and spur them to a full gallop. How I loved the sensation of speed, the rush of adrenaline. It was like flying. Like being free.

The boys probably didn't appreciate my zeal. After all, I was supposed to be submissive. And my father eventually told me—when I was eleven—that I should stop riding horses, that it wasn't ladylike.

That was a twist, certainly. We were made to look like hardy European peasants, but riding horses was a no-no.

I didn't argue, though. Not then. I found other ways to express my competitive fire, my individuality.

Even today, I remain a tomboy, with my alternately wavy,

frizzy auburn hair, cherubic, rosy-cheeked face, sturdy hips and thighs, and strong hands.

But I have come to appreciate a modicum of makeup—little accents of mascara, lipstick, and blush—and I will experiment with different attire. Colors I couldn't wear when I was Amish. Skirts that are a tad shorter than the ones I used to wear.

When *Glamour* magazine came to do a story on me after I'd left the farm, they spent two hours doing a makeover, arranging my hair, and putting more makeup on me than I'd ever had in my life.

I liked the pampering and I liked the new look, but I've never tried to duplicate it. If anything good has come from growing up Amish, it's that I appreciate what's on the inside of a person, not just what's on the outside.

Too much makeup can be unappealing. Likewise, I'm not a fan of women who bare too much of their bodies. A little left to the imagination is not a bad thing; sometimes when you don't expose everything, you're prettier. Like Catherine Zeta-Jones in *The Mask of Zorro* or Maureen O'Hara in *The Quiet Man*.

I can't say I'm a raving beauty as they are. But am I unattractive? No, I don't think so.

There's this saying that some people can be as ugly as a mud fence on the outside, but they can have beautiful personalities on the inside. And I'd like to think that I have a good personality.

Just like my mom. She has a good heart, and she would do anything for anybody. I don't think for a minute she would be writing what she does to me—saying what she says—if she weren't under the watchful eye of my father.

I think she once had a free spirit as I do, and I think she would have done so much more with her life if she could have.

Like me, one of the things she loves is traveling. But my dad rarely takes her. His idea of traveling is to visit other Amish—to boost his stature in social circles. My mother's idea of traveling is to sightsee, but my dad says that's a waste of time.

Mom once said she didn't want me to fall in the same rut she was in. Stuck and unhappy, a spirit in shreds. A life that plods along one day to the next. A life always on the edge of fear.

I hope one day she will come to know the glow that dwells within me—the smile I carry in my soul and wear on my face every day.

My mom doesn't smile much, but when she does, she's beautiful. People at my new church say I've got a smile that will melt a rock, and I guess in that way I'm beautiful, too.

In any event, I don't want to be anyone else. I'm perfectly content with what I am, and excited about who I will become.

Thanks, in no small part, to Ottie Garrett.

Five

Ottie, do you remember the times we visited on different trips we took? The one question that I ask you [is] if someone was in the Army if they could ever be kind again? For someone had told me that once someone is in the Army they take all the kindness away.

LETTER FROM MOM

Ottie Garrett rolled into Kalona in 1989, the year the Berlin Wall fell. He was among hundreds of English people across the country hired by the Amish as drivers for up to sixty-five cents a mile, and he was in town bringing relatives from Indiana and Illinois for Thanksgiving.

I was a naive fifteen-year-old who had recently graduated from the one-room Centerville School, a half-mile up the road from our farm. He was a forty-year-old man of the world, married three times, about to get another divorce, and more boisterous than anybody I'd ever encountered before or since.

He was the first person I'd seen who was unafraid to speak his mind among the Amish, and his views often challenged the community's befuddled elders. He was also playful and gregarious, traits both foreign and troublesome to the Amish.

Everything about him was big—from his 450-pound frame to his sleek, fifteen-passenger Ford van, complete with snazzy running lights, loud glass-pack mufflers, CB radio, and radar detector. And he loved to drive fast, with a Diet Coke in one hand and the wheel in the other.

My grandfather on my father's side, Tobias J. Miller, is the one responsible for Ottie staying on in Kalona. He asked Ottie during his visit if he'd like to be his driver. Over time, Ottie developed that initial offer into a thriving taxi service for my family and other Amish. He also began producing Amish calendars depicting their buggies and horses, and later several Amish-related books.

He became the talk of Kalona—from its tiny, throw-back business district of antique shops and hardware stores to its robust farms spread out along miles of hilly dirt roads. He stood out for his free spirit and humor. For his zest for life. For the magic tricks he'd do for the children.

I'd watch young children giggle with delight when he'd will an invisible ball to smack the inside of a paper bag, or make a cigarette vanish then reappear, or turn a single kernel of corn in his mouth into five or six.

I'd watch other Amish my age stare in amazement when he'd roar through town in that spiffy gray van of his, or maneuver it to do circles in the snow.

And to hear him talk was like nothing we'd heard before. He had a saying for anything, a bravado that was as large as he was.

One of my favorite stories is the time Ottie exited the van to let out an elderly female rider. When he opened the side doors and she stepped out, it was the first time the woman had seen him standing. She was startled by his imposing stature.

"My goodness," she said. "I didn't realize you were that big. What do you eat?"

"Little old Amish ladies like you," he replied, smiling mischievously. "And I haven't had my lunch yet."

The woman abruptly turned and walked away. Briskly.

Some people might say Ottie's got a line of bull a mile long—he might say so, too—but it all comes from a good heart. Some people also might say he's a bit unattractive, what with his stout girth and crippled right leg, fused stiff at the knee after a construction accident while working as an architect for the U.S. Department of Labor.

But I saw beneath his flaws. More to the point, I didn't see them at all.

To me, he was the most refreshing thing I'd encountered in my fifteen years. And although I didn't fully recognize it then, he was ample proof that not all English people were bad.

This was a concept I had always wanted to believe, despite the gloomy bias of the Amish that, early on, had me feeling sorry for the English and their supposedly depraved ways. I had wanted to believe otherwise because God's good graces told me there simply had to be good people in the outside world. Jesus, after all, had no hidden agenda. His was a ministry of sincere caring—for everyone, not just a select few. Not just for the Amish.

Psalm 33:13 says: "The Lord looketh from heaven; he beholdeth all the sons of men. From the place of his inhabitation he looketh upon all the inhabitants of the earth. He fashioneth their hearts alike; he considereth all their works."

God, without question, then, was an advocate of inclusion, not exclusion. And that meant Ottie Garrett, size and all, was one of God's children.

For two years, I dreamed of the day that I might ride in Ottie Garrett's van, to do more than merely glimpse this man of mirth. It was not a lustful desire; I didn't have such feelings. I just wanted to be a part of his joy, to savor his good nature and independence.

My chance came in the spring of 1991 when my grandfather planned a trip to Canada to see relatives and friends. Benedict and I were invited because we'd struck up a friendship with three Canadian children our age when they'd visited Kalona.

Benedict had begun corresponding with the boy, Danny, and I had exchanged letters with the girls, Ruth and Fannie May. Benedict was twenty and I was seventeen when we stepped into Ottie's van for the first time.

We sat in the back because we'd heard that was the best place to feel the power of a van. Besides, the grown-ups wanted to sit up front—to talk to Ottie and keep an eye on him. I could never figure out why they subjected themselves to such torture, because Ottie would invariably engage them in debate about their religious beliefs, and they would by turn wilt under the persuasiveness of his logic.

One of his biggest sticking points was the fact that the Amish choose to be conscientious objectors when it comes to serving in the armed forces. The Amish believe war is something to be detested. That's why they prohibit mustaches; they consider them a violent sign of masculinity, a symbol of the warrior, perhaps harkening to the days of Hitler's Third Reich or the Civil War.

The stubbornness of the Amish to take any responsibility for the freedom of their adopted country infuriated Ottie to no end, and he would pull no punches in taking them to task.

"So in other words," he once told an Amish bishop, "what you're saying is: You want everyone else to go die for you so you can sit back and tell us English how wrong we are."

"Well, the Constitution says we don't have to fight," the bishop replied.

"That Constitution wouldn't have been worth the paper it was written on if Hitler had gotten to us," Ottie fired back. "That piece of paper doesn't save you. It is the blood of other men that saves you. Just like the blood of Christ saved you."

It was at this point that the Amish elders usually became tongue-tied, determined to win the war of words but suddenly unable to articulate their position.

"Well, we're not going to fight," the bishop insisted.

"That's your constitutional right," Ottie said. "But at the same time I find it really rough that you people look down on us, condemn us to hell, and tell us how bad we are—but you live by our Constitution.

"I have a question. How do you feel about the woman down the street whose son has just been killed in the war defending your son?"

"They're not Amish and we're not English," the bishop said. "We're from two separate worlds."

On another occasion, smack-dab in the middle of Desert Storm, an Amish bishop told Ottie: "Because of the war, we have grown in strength because our young men have seen the light and come back to the church."

"No," Ottie said. "They've come back because they're cowards."

Ottie, a former U.S. Army infantry soldier who served eighteen months in Wildflecken, Germany, would later tell me such conversations diminished his respect for the Amish.

"When they have Veterans Day, the Amish could care less," he would complain. "They don't even honor the flag, or all the men who sacrificed for them."

This, of course, was the Ottie I didn't yet know as we headed north to Aylmer, Ontario. What caught my attention during that first drive were things far less weighty.

The speed with which he drove, sometimes as fast as eighty-five miles an hour, appealed to my thirst for adrenaline and competitiveness. It beat the pants off our horses and buggies, which clip-clopped along at a leisurely five to eight miles an hour. And it was thrilling when we'd pass other cars, rather than them passing us.

The music he played on the van's tape deck was another eye-opener. The words—in this case those of the Statler Brothers—were accompanied by instruments.

In the Amish church, everything is sung a cappella. Ottie says my singing voice sounds like an angel, but this addition of instruments was glorious—and in keeping with my impression that God intended for music to be grandiose.

The Book of Psalms, especially those of David, is set to the accompaniment of stringed instruments and is intended mostly to be a joyful praise of God's mercy, not a solemn processional. It is the original biblical hymn book.

Psalm 33:1 says: "Rejoice in the Lord, O ye righteous: for praise is comely for the upright. Praise the Lord with harp: sing unto him with the psaltery and an instrument of ten strings."

I remember sitting in the van, listening to the Statlers, and thinking: This is the way it's supposed to be. This is how God should be worshiped.

I also remember Ottie's knack for spouting phrases like "putting the pedal to the metal" or this ride "will be smooth as a ribbon." They sounded suave and highly intelligent to ears that had never heard such things before.

I didn't mention my observations to Ottie on that trip, preferring instead to keep my newfound knowledge to myself. And Ottie said nary a word to me. He thought of me as just a kid, and so I was.

But I had gotten to ride in Ottie Garrett's van. For the first time, I had shared space with this marvelous man.

Six

Irene . . . [there] still is a wrong and right way to live. We pray for you many times. Also, still burn a light at night. Again, I apologize for any cause in words or actions for your going, and pray to be forgiven by our creator and Savior.

—LETTER FROM DAD

The year before the Canada trip with Ottie, I had taken one of the most significant steps an Amish person can take. I had become a member of the Amish church.

Formally joining, which usually occurs after a person turns sixteen, represents someone's desire to make a lifelong commitment to the Amish church—and it is a process not taken lightly. In fact, people are commanded to join, and run the risk of being shunned if they don't comply.

Prospective members must attend nine Sunday classes—instructions of baptism—that review the eighteen articles of the Mennonite Confession of Faith, drafted in Holland in 1632.

They begin: "Whereas it is declared that 'without faith it is impossible to please God' (Hebrews 11:6), and that 'he that cometh to God must believe that he is, and that he is the rewarder of them that diligently seek him,' therefore we confess with the mouth, and believe with the heart, together with all the pious, according to the Holy Scriptures, that there is one eternal, almighty, and incomprehensible God, Father, Son, and the Holy Ghost, and none more and none other, before whom no God existed, neither will exist after him. For from him, through him, and in him are all things."

The document covers sundry subjects such as "The Fall of Man"; "The Restoration of Man through the Promise of the Coming Christ"; "The Advent of Christ into This World, and the Reason for His Coming"; "The Law of Christ, Which Is the Holy Gospel, or the New Testament"; "Repentance and Amendment of Life"; "The Washing of the Saints' Feet"; "Matrimony"; "Excommunication or Expulsion from the Church"; "The Shunning of Those Who Are Expelled"; and lastly, "The Resurrection of the Dead and the Last Judgment."

It ends: "May the Lord through his grace make us all fit and worthy, that no such calamity may befall us; but that we may be diligent, and so take heed to ourselves, that we may be found of him in peace, without spot, and blameless, Amen."

After the ninth lesson, we got on our knees before the minister's bench at the front of the church and were asked a series of questions, each having only one correct answer. They were asked—and answered—in German:

Q. Can you confess with the eunich, "I believe that Jesus Christ is the Son of God?"

A. Yes, I believe that Jesus Christ is the Son of God.

Q. Do you also confess this to be a Christian doctrine, church, and brotherhood to which you are about to submit?

A. Yes.

Q. Do you now renounce the world, the devil, and his

evil lusts, as well as your own flesh and blood, and desire to serve only Jesus Christ, who died on the cross for you?

A. Yes.

Q. Do you promise, in the presence of God and his church, with the Lord's help to support these doctrines and regulations, to earnestly fill your place in the church, to help counsel and labor, and not to depart from the same, come what may?

A. Yes.

After we correctly answered the questions, the bishop's wife removed the girls' head coverings. As the bishop cupped water in his hands above our heads, the bishop recited the baptism: "I now baptize you in the name of the Father and the Son and the Holy Ghost. . . ." Water streamed everywhere when the bishop's hands parted. Down our necks, our ears, and our faces. It was more than a drop or two. It was a shower.

The ceremony concluded with holy kisses from the bishop (for the boys) and from his wife (for the girls). The kisses were on the cheeks, though that would later change for some of the boys and girls. Bishops, ministers, deacons, and married men shake hands and kiss each other on the lips upon arriving at church. The women do the same among themselves. The practice essentially comes from the Bible, Thessalonians 5:26: "Greet all the brethren with a holy kiss."

My joining the church came at a critical point in my life, for I had begun to notice other contradictions in the Amish way that didn't seem fair. And that's exactly how a teenager, Amish or otherwise, might describe them. Not fair.

Some post-school Amish teenagers—toughies or fence-crowders, we'd call them—would get away with alterations in their dress that, no matter how slight, nevertheless violated codes. Why, I thought, are the rules enforced for one person but not another?

And, although we were prohibited from owning phones and cars, we used them regularly. Phones were conveniently placed in barns or cloaked in shacks behind trees. Hired drivers like Ottie, meanwhile, traveled more than 100,000 miles a year, transporting Amish to friends, relatives, national steering committee meetings, or far-flung vacations.

It also seemed peculiar—not fair again—that it was okay to leave the Amish for more liberal orders like the Beachy Amish or the Mennonites, but it was frowned upon to depart to other denominations.

This philosophical dichotomy was evident just down the road from our farm. Buggies and horses would fill the driveway and yard around a house hosting Old Order Amish church services. Half a mile away, cars would fill the parking lot of a Mennonite church.

Which way was truly right? And if it was okay to morph into a Mennonite—and therefore own cars, wear more colorful dress, and let one's hair down—why couldn't a person become a Baptist, a Lutheran, or a Methodist without facing the threat of excommunication?

The Amish, pacifists to the end, will tell you the other denominations are warring churches; that they allow parishioners to engage in armed combat. And that makes them unsuitable houses of worship for Amish defectors.

But my sense was that the parishioners in the other churches believed in God just as much as I did. They prayed to the same God. They read the same scriptures, for the most part. They had the same desire to give their lives to Christ.

I also wondered, if the Amish were so privileged, if they alone had the key to God's graces, why were they so adamant about not ministering to other faiths? Wouldn't they, in good Christian conscience, want all of mankind to know about the route to salvation? Wouldn't they want to share their good fortune?

The Bible instructs Christians to minister afar in Mark 16:15: "And he said unto them, Go ye into all the world and preach the gospel to every creature."

The Amish church didn't take long to feed my doubts. The first confessional involved my brother Elson. The Amish require that rubber wheels be taken off tractors and replaced with impeding spiked steel wheels—so people aren't tempted to use the tractors like cars. My brother had used the rubber wheels just once, but it was enough to require he confess. His brother-in-law, meanwhile, was a regular user of rubber wheels and had never been punished. I guess they were trying to make an example of Elson, but it didn't seem right.

As with my other concerns, I didn't mention any of these to anyone. I didn't dare. Thinking aloud too much, analyzing too much, would raise suspicions. And that would make life hard, if not unbearable. The Amish are always watching to see if a person appears vulnerable to doubt. It is their greatest fear.

The flock must be strong and unified to survive, they believe, and they often mention Noah tarring the ark to keep water from entering. Similarly, they say, the Amish church should be tarred inside and out to keep the world from seeping in.

The church and its inhabitants must never waver.

And so I didn't. At least not publicly.

But my growing, private misgivings would play a big role later—when everything was on the line.

The key distinction I made when I joined the church was that I had committed more to God than institution. To me, that was the purist route to faith.

Seven

Why, oh why, did you walk out on us like this, without a goodbye or a talk? The schoolchildren say Irene said she would show us how to make a kite and now where is she? . . . There are so many dark shadows you have cast over so many lives.

—Letter from Mom

I didn't stop thinking about Ottie after the Canada trip, but I didn't see much of him for the next three years.

Occasionally, I'd glimpse him as his van zipped down the road, and I'd hear Amish talking about him from time to time.

Mostly, though, my life was a mundane routine of work and sleep. I'd help Mom with the gardening, the canning, the mowing, laundry, and sewing. During hay season, I'd drive the tractor with the bailer behind it and the men would stack the flaxen stalks.

All the while, I was counting the years until I turned twenty-one, when I would be able to travel more freely, when I could see the world beyond the little shops, big farms, and manure-tinged air of Kalona. The Amish permit a person more personal choices after they turn twenty-one.

As luck would have it, my opportunity came earlier than that. In the dawn of 1994, Ottie began coming to our farm for foot treatments from my mother, a recognized reflexologist with a steady Amish clientele. Her massages would help ease Ottie's gout and tendinitis.

His visits were infrequently regular, if there is such a thing. Once or twice a week. Once every two weeks, sometimes. All hinging on when he'd be in from the road.

Having Ottie around once again filled my life with a joy, a freshness, that was intoxicating. It seemed so impressive that he was coming to our house.

I'd make any excuse to walk through the room where Mom was doing her reflexology, and Ottie and I would exchange pleasantries. Nothing remotely intimate, of course. Just friendly conversation.

But I found the more he visited, the more I missed him when he was gone. It was an ache completely foreign to me. Although I was now twenty, I still had not dated, and I had only the innocent notion that a feeling like this must be the sign of a blossoming friendship. A platonic closeness.

A couple of other interesting things accompanied the yearning. Ottie had begun taking pictures of Amish buggies—a hobby that he had turned into the business of making calendars. He'd also sneak in a few pictures of the Amish themselves.

He had to be adept. The Amish view taking pictures of people as something akin to making graven images: "Thou shalt not make unto thee any graven image, or any likeness of any thing that is in heaven above, or that is in the earth beneath, or that is in the water under the earth" (Exodus 20:4).

As such, the Amish will not pose for pictures, and they often adopt a look of disgust when they know their pictures are being taken. Further, businesses in heavily populated Amish areas often remind people not to take pictures of the Amish. One such place was the Stringtown Grocery in Kalona, where patrons could buy everything from muffin and bread mix to all manner of sugars and candies. "No cameras allowed," a sign on the grocery's front door reads.

But Ottie would find a way around such restrictions. He'd shoot the Amish outright when they weren't looking. Or, he'd use a long lens, tell the Amish he was shooting their black buggies, then twist ever so slightly until the Amish came into frame.

I knew when he was taking my picture. I'd try to look disapproving, but occasionally I'd smile. Sometimes right at him. Chalk it up to another Amish contradiction—while many appear as though they dislike being photographed, inside there is a fascination with the process. Some Amish even enjoy watching English people stumble over themselves trying not to take a picture.

Ottie's photographing became important when I and other family members began taking more trips with Ottie behind the wheel. To Canada again, Missouri, Indiana, Ohio, Virginia, Tennessee, and the Smoky Mountains, a favorite of mine. And much later on, Florida, when my life began unraveling at blinding speed.

Traveling was the ultimate in learning. It gave me a chance to see things I'd only read about, sometimes things I'd never even heard about.

Ottie's photographic images from these trips, meanwhile, are tucked in photo albums that to this day allow me to enjoy, time and again, the glories of travel—and to look back on how I once lived.

Our trips brought me even closer to Ottie.

I had already begun a habit of baking him oatmeal-raisin cookies, or apple bars, or pumpkin bread, or cinnamon rolls whenever he'd visit Mom for treatments.

On the surface, they were harmless gestures. He was, after all, the family's driver, someone to be nice to.

But we both began to believe, separately at first, that perhaps our growing rapport might be something more. For my part, I knew that nothing could come of it—for several reasons.

The Amish believe that if a member of the flock marries a divorced person while their ex-spouses are still alive, it's adultery. Beyond that, marrying an English person would mean leaving the Amish, and both that and the adultery issue would mean being put in the ban. I would not be able to return unless I promised never to marry again.

Certainly, some people would also condemn a relationship with a man twice my age. I was now twenty-one, he forty-five.

I also felt obligated to stay and care for my mom. To protect her from my father. To be there when her health wavered.

In the beginning, I didn't spend much time contemplating the possibilities of Ottie and me, because anything more than a friendship simply wasn't possible. Ottie fostered the same understanding, although he increasingly found it hard to erase one particular image from his memory.

He had come to pick up my mother one sunny fall morning and had noticed me hoeing a flower bed in the front yard. I was barefoot and had on a long, dark blue dress—sleeves rolled to the elbows—and a light blue scarf on my head. He would later tell me the sun seemed to be bathing me in a halo of gold and, when a light breezed kicked up, it tugged at my dress, showing off my hips, legs, and waist.

"Hello, Ottie," I said, walking to his van and brushing aside a tuft of hair that had escaped the scarf and landed on my brow.

"Hi, Irene," he said. "I wish you could be going with us."

"I wish I could, too, but I've got to take care of the house while Mom's gone."

Ottie nodded, caught up in my smile, and for a moment time seemed to stand still. Then I remembered I had forgotten something. Ottie said he watched me, mesmerized by my gaiety, as I ran inside the house and returned, bounding off the porch in one leap, with a plate of homemade cookies in my hands.

He said later that's when he knew that, beyond any physical attraction, he was being drawn to me by my independence, my mental strength, and my radiant spirit. And I had begun to realize that, beyond Ottie's charm, I was being drawn to him by a kindness and caring I'd never known before.

I handed him the plate of cookies and, as I did, our hands touched ever so slightly, ever so briefly. We didn't acknowledge to each other—or even to ourselves—that this coincidental alliance was invigorating, but in hindsight it was.

The next time it happened, the circumstances would be altogether different.

On one of our summer trips—to visit relatives and friends across Indiana—Ottie arranged for me to be the last one let off, a rare circumstance since one of my brothers—or Bertha, my older sister—was always around to act as a chaperone. I was, after all, a single woman.

Ottie pulled the van to the side of northeast Indiana's Highway 20, between Shipshewana and Middlebury, and came straight out with it.

"Look, Irene," he said, "we've got a problem. My feelings for you are growing."

"I feel the same way, but I could never leave the Amish," I said.

"I know. And I'm not supposed to feel this way. You're so young and I'm so old."

"I don't care about that."

We looked at each other for a while, absorbed, suspended in a passionate gaze. Then he held my hand. Firmly.

"Anything I can do for you, I will," Ottie said.

Currents of excitement—and fear—charged through me. I couldn't help but wonder if it was okay to feel this way. Was it a sin? I wondered. Would I have to openly confess in church? I hadn't really done anything. Or had I? Wasn't the issue what I felt inside, not what I had done?

Worst of all, I wondered, had I made myself vulnerable to the situation by more fervently embracing my discord with the Amish way? Shortly after I turned twenty-one, I had purchased a fifty-five-dollar harmonica, the only instrument allowed by my community. The reasons for the exception are unclear to me, but my motivations for buying the mouth organ were unequivocal. I loved music, and I was lashing out at my lack of freedoms.

Owning the harmonica was a guilty pleasure I could

accept, and rationalize. Grasping Ottie's hand was another matter, and it sent me spiraling into a maze of confusion. This was not supposed to be happening—yet it was.

I had held hands with someone for the first time, violating fundamental Amish dating etiquette. To an outsider, it doesn't sound like much. But for me, my whole being had been turned inside out.

The first place I turned to when we got back to Kalona was the Bible.

Psalm 56:3–4: "What time I am afraid, I will trust in thee. In God I will praise his word, in God I have put my trust; I will not fear what flesh can do unto me."

Isaiah 41:13: "For I the Lord thy God will hold thy right hand, saying unto thee, Fear not; I will help thee."

If I could just hold God's hand long enough, I thought, everything would be all right. And for a time, it was.

Ottie and I managed to keep our emotions in check for the rest of the summer, and the leaves of fall brought a momentary diversion.

I started teaching school.

Every morning, I'd climb into my buggy—I had to hold on tight because it would list until I was in place—and head the three miles to the one-room Shady Lane School.

It was pretty typical for an Amish school, and similar to the one I had attended as a child. A white, wood-sided structure with a concrete foundation painted sea-blue. A little bell housing on the roof. Old, lift-top desks and chairs attached by a metal frame. A mammoth tree stump, a slide, swings, and a merry-go-round in a side yard. A ball field out back with short fences and a wood and chicken-wire backstop. A silver outdoor pump that took a dozen or so hard pushes and pulls to draw drinking water.

I had fifteen children in my classroom, and was paid eighteen dollars a day for my efforts—or about $2.40 an hour. Minimum wage standards did not—and still do not—extend to the Amish.

I patterned my teaching methods after a favorite instructor of mine at Centerville School. She was a robust, fun-loving woman who enjoyed children and upheld rules with a firm but compassionate touch. She was my teacher for the third through eighth grades, and she was a far cry from my first instructor.

I will name neither here, but my first one was a skinny, bony-fingered sadist who spoke with a lisp and delighted in punishing children to extremes.

There were a number of things that could land a student in trouble. Dropping books on the floor, because it made too much noise. Talking out of turn, because it was disrespectful. Looking or smiling at other students, because it suggested a lack of attention to the lessons. Cheating and passing notes, because what school doesn't discourage those?

I remember when that first teacher punished two friends of mine for passing notes by having them clean a toilet with bare hands and a rag. A brush was available, but she forbade its use.

Other times, she would punish students by having them stand in a corner with their noses pressed against the wall. And occasionally, she would draw a circle on the chalkboard and have the errant students stand before the board with their snouts in the bull's-eye. For ten minutes. Maybe fifteen.

It was utterly humiliating. I know, because even though I was a good student and a respectable child, I had my fair share of discipline. Once, I remember, I made the mistake of looking too long at another pupil in class. I wound up with my back to my colleagues and my nose in the circle.

That first teacher also dished out spankings with a wooden paddle. They were done out of sight of the rest of the class, probably after school, when she did a lot of her disciplining—just to inconvenience the accused, I suppose.

I don't think there was a child under her tutelage who liked the young woman, and I don't think there was one of us who was unhappy when the second teacher came along.

The second teacher also paddled children, but she was more yielding with her most common form of punishment: sweeping the floor during the third recess. And if you got done before the recess was over, you could go outside and join the other children.

As it turned out, I adopted the same reasonable discipline for my students. I also carried forth with an attitude of joy and caring, and borrowed from my second teacher such activities as putting positive messages or treats in plastic eggs and hiding them around the school property.

I even invited Ottie once to teach a Friday class in drawing, something I'd seen him do earlier when he decorated the inside of Aaron's hymnal with an eagle or drew pictures of places he'd visited in his travels. In class, he drew horses, buggies, cowboys, and cartoon characters. He also taught the children how to construct a person's face by drawing an egg shape, splitting it into quadrants and starting the eyes just below the center horizontal line.

The children were enthralled by such unbridled creativity.

I vowed I would never paddle a child, and in truth, I never needed to. When a teacher treats students with respect and admiration, the students usually respond in kind. At least they do in Amish schools. At least they did with me.

Teaching school gave me a sense of purpose and less time to think about my growing affection for Ottie. But in my weaker moments, I would let my guard down and my mind would wander.

Ottie didn't make it any easier. He began bringing me gifts from the road, including beautiful—and expensive—crystal swans.

I'd hide the swans in my chest of drawers; certainly, my parents would consider them inappropriate gifts. But just like my feelings for Ottie, I knew where the swans were.

They were never more than an inch or two from my heart.

Eight

Ottie, how could you do this to us? . . . You took her so she could not keep her promise to teach school again and her promise to the church on bended knees. We took you in as a trusted friend, tried to help you in time of sickness, and trusted you as a friend. Now you proved yourself not worthy of the trust at the cost of our darling daughter Irene.

—Letter from Mom

Kalona is the kind of place where everyone knows everyone else's business and, even if they don't, they want to. It's a place where neighbors know who the owner is when a dog barks, where Midwestern values—and provincialism—are protected fiercely.

It's also a tourist landing where thousands alight each year to see the Amish farms on the town's outskirts—and the more modest Amish dwellings within the city limits, where horses graze in the fenced backyards of some homes.

People come for the three-day quilt show and sale in late April, the Kalona Fall Festival in September, and the Kalona Historical Village, a collection of restored nineteenth-century buildings. Every Monday, they venture to the Kalona Sales Barn, where the proprietors offer horses, cows, sheep, and the like.

There are the downtown staples like Reif's Family Center, Yotty's Hardware, and the weekly *Kalona News,* and the quaint-sounding businesses like Miller's Medicine Cabinet, the Wooden Wheel quilt shop, and Ellen's Sewing Center. A stone's throw from the sales barn is Kalona Blacksmith & Welding, where the owner hangs a metal sign by the front door when he leaves. "Out on call," it says.

Just outside town, people can buy curds and such from the Kalona Cheese Factory, which proudly claims it will ship anywhere.

Hills Bank has a diminutive white clock tower that nevertheless is the tallest structure in town. If that's not enough to keep people on time, an air-raid siren goes off at noon every day. People set their clocks by it, especially the Amish, whose timepieces are powered by batteries or pendulum.

Visitors are just as apt to see an Amish buggy and horse affixed to a hitching rail as they are a car parked in one of the downtown's diagonal spaces. In Kalona, the natives like to say, the English and Amish coexist harmoniously, one living in the twenty-first century, the other a hundred or more years in the past.

Residents call it the heart of Iowa's Amish country, and boast that the seven hundred Amish inhabitants make up the largest such settlement west of the Mississippi. The town also goes by another moniker: "Quilt Capital of Iowa."

The Amish were the first to settle in Kalona, arriving along the banks of the English River (oddly enough) in 1846. The area was nameless then. In 1879, it became Bulltown, after a successful shorthorn breeding service. Later, it became Kalona, the name of the service's famous registered sire.

It was a truly bucolic, out-of-the-way burb until the 1950s, when Highway 1 was paved, providing easier access to and from Iowa City. Ottie says construction crews used creek gravel in the pavement mix; hence, the unusual pink hue.

The highway brought more English—some from foreign continents—to Kalona, and some people stayed. Kalona's growth challenged its pastoral ambience, but the larger the town grew—to more than 2,000, by some estimates—the more determined it became to preserve its heritage.

The town's original motto speaks to its desire to meld tradition and progress: "Big enough to serve you, small enough to know you."

It was in this fish bowl that Ottie and I existed. Known by everyone. Watched by the English and Amish alike.

We developed a system of glances that would let the other know we were thinking of them. Ottie, meanwhile, kept bringing gifts. Flowers sometimes, or chocolate truffles that he would cleverly share with the rest of the family.

I continued baking him goodies.

Ottie, formally separated from his wife for eight months now, had moved into a two-bedroom, gray bungalow on Kiwi Avenue off Highway 1. The house, which he rented for five hundred dollars a month, stood alone, surrounded by corn fields and a spotting of poplar and pine.

He arranged to have me work for him weekends, tidying his house, dusting and cleaning, doing paperwork, and tending to the garden. A mountainous man with a cane doesn't get around easily.

He also hired my sister Bertha and several of my brothers to mow the lawn and paint the fence, although they were more my chaperones than Ottie's employees.

The best part of the arrangement was being paid a princely sum of five dollars an hour, ostensibly to be closer to Ottie. The worst part was the temptation.

And this time, it was me who took the lead. In February 1996, I stopped to do some work at his house—and to secretly leave him a present. I had cooked a container of popcorn and included with it several Hershey's Kisses. It was my unspoken way of hinting I wanted a kiss.

The next weekend, Ottie asked me about the chocolate drops. When he saw my face flush, he knew.

While Benedict shoveled snow out front, Ottie moved toward me, softly kissed my lips, and stood back, waiting for a reaction—one he would never see. Although my insides were instantly consumed by a fluttering giddiness, I was stiff as a board on the outside, and for him it must have been like kissing a rock. I didn't move. I didn't close my eyes. I didn't even open my mouth.

When you've never been kissed before, you don't know how to react. You don't know about tongues, and opened mouths, and sharing saliva. In my case, the only comparisons were the holy kisses among the men and women at church. The tight-lipped holy kisses.

Nevertheless, the moment broke another barrier, and our complex series of clandestine signals grew by one. Added to the repertoire of knowing winks and smiles was a tug on the ear lobe. If either of us did it, it meant we wanted a kiss.

We also began teasing each other behind my parents' backs—and in the process threatening to have them, and others, discover us. Ottie would occasionally tweak my behind when he walked by, startling me enough to make me jump.

I was just as game. Once, when my mother and father were in their bedroom, I walked up to him in the kitchen and planted a firm kiss on his cheek. He later told me: "All I could think about was Alvin coming around the corner and saying, 'What's this!' And what would I do? I can't run."

Another time, while we were eating dinner at my family's house, I began massaging his foot with mine under the table. Ottie was at one end of the table and my father at the other and they were engaged in conversation. Ottie became so flustered that he began stammering. Later he would tell me: "Irene, that wasn't funny!"

But it was. For both of us.

It was also exciting. And frightening again.

I began praying more in private, occasionally in my closet. I had read that if one does that, God will reward you.

Matthew 6:5–6: "And when thou prayest, thou shalt not be as the hypocrites are: for they love to pray standing in the synagogues and in the corners of the streets, that they may be seen of men. Verily I say unto you, they have their reward.

"But thou, when thou prayest, enter into thy closet, and when thou hast shut thy door, pray to thy Father which is in secret; and thy Father which seeth in secret shall reward thee openly."

Always, I would ask God to lead me in the right direction. Often, I would cry because I was filled with so much confusion. How would I know if it was God or the devil that was taking me down this path? How would I ever sort out the scriptures that seemed to both doom and support such a relationship?

Putting my faith in God allowed me some breathing room. But at the same time, the relationship grew deeper.

We began slipping each other notes, putting into writing what we had previously only said or signaled or acted upon.

I put one note in a container of banana bread I had left in our oversized silver mailbox. Ottie was preparing for another trip, and I told him he should stop by the mailbox when he headed out of town around 2:00 A.M.

Dear Ottie:

Please don't let anyone see this note. Here is your banana bread and I'm so sorry we couldn't persuade dad to go along.

It hurts me, but remember, I'm going to miss you and think about you every day. I really hope you have good luck to make this trip worthwhile for you.

Ottie, I'm going to be awake when you pick up the banana bread. If no one is along, you could whistle and I'll whistle back. (To say "bye.")

How I would want to meet you by the mailbox and talk, but I was afraid I'd be heard going out, so I thought it better I wouldn't.

Don't whistle if you don't want me to whistle. But I'll be watching.

Love, Irene

Ottie later slipped me this note:

My Dearest Love,

My heart aches to be with you. My arms want and need to hold you close and take care of you.

I wish nothing but to hold and comfort you and take care of you all the time. My every thought is of you and only you.

I can only pray that some day I'll be able to do all of these things for you.

In dreams, all my love and kisses,

Yours truly,

OG

His mention of "in dreams" referred to a Roy Orbison melody that had become our song. The melancholy ballad tells of a star-crossed couple who can only be together in dreams. Like us, pretty much.

Whenever we were in the van, Ottie would play the song. None of the other passengers knew its significance. But we did.

Sometimes, Ottie would also give me cute little cards with short messages. One said on the front: "Next time you're admiring all the wonderful things God has made . . ." It continued inside: "Remember that you're one of them." Another had a touching picture of two swans floating breast to breast. Ottie wrote on that one: "Sometimes there are no words."

The first time Ottie told me he loved me, I thought: Why would anyone feel the need to say anything like that? Love, after all, is not used in that context among the Amish.

The second time he told me he loved me, he proposed.

"Let's leave," he said. "Let's get married."

"I can't," I told him, "because I'm Amish. I can't because I can't leave my mother or the rest of my family."

"Okay," he said, joking. "I'll stay here in Iowa and wait for you. I'll wait until you're twenty-five, and then I'll have to kidnap you."

"You don't have to kidnap me," I said.

And we both laughed.

But privately, I was in pieces.

I had written Ottie before about my conflicting emotions, so my reluctance was no surprise. It must have been frustrating for him, though.

My Dearest One:

I wonder what you're doing tonight with these many miles between us. I'm crying my heart out for you. I don't know what else to do except cry and pray.

This morning when you were here, you seemed so sad. I felt so sorry for you, how things are going, that I wanted to put my arms around you and comfort you. I love you so much, Ottie! I know how you feel and, oh, it feels like my heart is being ripped out. To think of you leaving and I can't go with you is almost more than I can bear.

What am I supposed to do? How can I leave Mom when I know she will be blamed, abused, scolded, and maybe even hurt because of me.

There was one time when he was so angry, I became uneasy that he might seriously harm her. Later, I asked her if she's afraid that would happen sometime. She admitted she is afraid and cannot sleep if he walks "stealthy-like" through the house when he's so angry. She cannot sleep until he is also in bed and she knows he is sleeping. That scares me. . . .

I'll sort of have to see how things go to know what I'll do. But please remember, my love for you will never die! The red rose I pinned on your shirt will never fade—so is my love for you! I love you, Ottie!

Hugs and kisses,
Lovingly yours,
Irene

What I didn't know then—but do now—is that I'd already committed to Ottie. I was caught in a vortex of passion from which there was no return.

Nine

Irene, you vowed in your baptism before God and the church that you would be a building and an uplifting church member. Are you helping build up the Amish church? Are you being a help to the younger generation?

—Letter from Benedict

My grandfather, the one who had hired Ottie, died of pneumonia in May 1996. He had been one of the most powerful Amish men in Kalona, and not so much because he'd been a minister and a national Amish steering committee member. His truly important possessions were lots of money and all of the contiguous land east and south of town.

When he'd go to the bank, people would joke that T. J. was lending money to the institution. Sometimes, money and land holdings speak more than position among the Amish.

I liked my grandfather well enough, although I never felt particularly close to him. I remember he used to call me Ruthy and tease me a lot, especially about boys and such. But I don't remember us ever having meaningful conversation, and he was stern and unforgiving like my father.

He was buried alongside my brother and sister—Tobias and Miriam—beneath a stone almost twice the size of theirs. "Tobias J. Miller," it reads. "April 23, 1913, May 18, 1996, 83 Y., 25 D, Gone But Not Forgotten."

I wrote a poem to be read at his funeral. It was more for the benefit of others than for myself:

Dear husband, father, grandfather,
It is hard to see you go,
It was hard to see you linger,
And to see you suffer so.

Father, we have many memories,
How you helped in work and play,
Precious memories always linger,
Of our happy childhood days.

You have helped so many people,
Always willing, a hand to lend,
You have labored hard and long,
For family, church, and many a friend.

Jesus saw you growing weary,
So he sent an angel bright,
Softly whispered, "Come with me,"
To the realms of pure delight.

Dear grandfather, how we miss you,
And our hearts are sore with grief,
But we trust that you are resting,
In his arms of sweet relief.

Rest forever, rest in peace,
For your work on earth is done,
May we all meet you in heaven,
Where there is no setting sun.

My family chose not to use the poem at the services, picking instead some other published passages. And that was okay with me.

Although I felt sadness that day, his death was more a benchmark of my roller-coaster life. Things were moving so fast I could scarcely breathe.

Ottie's divorce had become final in April 1996, removing the stigma of being attracted to a married man. But when the notice appeared in the newspaper, people within the Amish community became suspicious about his intentions.

My father told me I'd have to stop working at Ottie's house. "That's gotta quit," he said, scowling. My deacon vowed to talk to me if I didn't comply. Meanwhile, Amish who had been using Ottie as a driver began telling him they no longer needed his services.

My oldest brother, Elson, also planned to intervene.

"I was going to talk to Irene about the connection they had," he would later say. "I saw some love there that was not supposed to be—a young girl like that falling in love with a guy that age, you know."

One of my uncles even got into the act. He spread information around the Amish community that he had seen Ottie and me look at each other in an intimate way.

Truth be told, it was more than just the divorce feeding the frenzy. It was us.

When Ottie was out of town, I'd use a phone tucked in a barn along the way to school to dial the motel numbers he'd left for me.

When he was in from the road, we'd meet before school—he in the van, me in the buggy. We'd pass along Johnson-Washington County Road about 7:00 A.M. and stop at a little dip where we thought no one could see us. We'd talk briefly, then move on.

Sometimes, if we mustered enough courage, we'd also meet at the school before the children arrived at 8:00 A.M.

Our after-school rendezvous were riskier still. I'd drive the buggy into town and hitch it inside a white shed off A Place, little more than an alley between 4th and 5th streets. Ottie would come down A Place in his van about 4:30 P.M. and I'd jump in the back and lie down until we were out of town and out of sight.

We'd drive south, traverse the English River on an old, one-lane iron bridge, and find a place to park in the country, where we'd talk mostly. About our separate lives. About Ottie's travels.

One of my strongest memories is the time he described driving west on the flatlands of eastern Colorado and seeing what looked like a long, gauzy band of clouds on the horizon.

"But what they were, when you got closer, were the Rocky Mountains—the most spectacular range of peaks in North America," he said. "Snow-capped sentries that seem like they go on forever."

"I would love to see them one day," I said.

"I would love to be there with you when you do."

On the days when we met up at A Place, I'd tell my family I was late getting home because I'd gone downtown to buy school supplies. Notebooks, paper, pencils, and the like.

That meant we had to time everything meticulously. We figured forty-five minutes was the most we could spend together before Ottie drove back to town and dropped me off. At 5:15 P.M., I'd run into Reif's, buy the supplies, jump in my buggy, and head home.

There was the matter of getting it all done before the stores—and the town—closed at 6:00 P.M. More important, dallying any longer would've surely raised suspicions.

As it turned out, we were not as careful as we'd thought. Word began filtering through the community that we had been seen once too often chatting in the middle of the road or conversing at the school.

Against this backdrop of intense surveillance, things were not getting any better at home. My father was executor of my grandfather's estate, and anytime my mother would ask him about it, she would be reprimanded. The tension that pervaded our house was unbearable.

Fortunately, a long-planned vacation was about to provide some relief—and then some. The week after my grandfather died, Bertha, I, and a cousin and her husband acting as chaperones set out with Ottie for a one-week trip that was to take us to Tennessee, Virginia, and Ohio. But when we got to Tennessee, Ottie, knowing I'd always longed to see palm trees and clear blue tropical waters, mentioned Florida was within reach. So off we went to Key West.

We drove Duval Street from one end to the other, over and over, watching the drunks, the transvestites, the sidewalk artists, and the motley-attired crowds. When we tired of that, we traveled the fragrant side streets of small homes, scampering geckos and lovers kissing under giant, twisted tree limbs.

Many Amish might have considered it hell on earth, and it certainly was a side of the English world I couldn't have imagined in my wildest dreams. But the energy of the place was fascinating. And I was away from Kalona—and my father.

Later that night, after discovering we couldn't get a motel room in Key West on Memorial Day weekend, Ottie parked by the beach and we slept in the van until the big southern sun began its climb over the Atlantic. Ottie and I went for a walk on the beach that morning and watched the seagulls hover like helicopters in the breeze. We didn't hold hands, because Bertha and my cousin and her husband were watching from the van. We didn't even look at each other when we talked, lest someone should conclude we were acting too chummy.

But the stroll was nevertheless romantic, and Ottie told me it could be this way all the time—only better.

I knew what he meant.

On our way back from Key West, we stopped in Berlin, Ohio. Bertha, who like Ottie had always had trouble with her feet—arthritis, I think—wanted to see Dr. John, an Amish doctor who'd developed a good reputation for treating such ailments.

She had initially planned to stay only a day or two, but decided to stretch it to two weeks when she secured lodging with an Amish girl in nearby Sugar Creek. Ottie then arranged for Bertha to work during her visit at a printing company he had done business with.

It looked like Ottie and I would finally be able to spend some unsupervised, unhurried time alone.

Ottie had booked three rooms at the Berlin Village Inn. One for me, one for him, and one for my cousin and her husband. Bertha would have shared my room if she hadn't extended her stay.

Ottie told me ahead of time that if I didn't want to come down to his room that night, he'd understand.

"But if you do come down," he said, "you're mine and you're staying the night."

For the first time, I was free of apprehension. I had already asked for God's forgiveness—many times. And I was secure in the knowledge that I had amply demonstrated my trust in him.

Psalm 32:7–10: "Thou art my hiding place; thou shalt preserve me from trouble; thou shalt compass me about with songs of deliverance. I will instruct thee and teach thee in the way thou shalt go: I will guide thee with mine eye. Be ye not as the horse or as the mule, which have no understanding: whose mouth must be held in with bit and bridle, lest they come near unto thee. Many sorrows shall be to the wicked: but he that trusteth in the Lord, mercy shall compass him about."

Maybe, I told God, what I'm about to do is a sin. Maybe, under any other circumstance, it is wicked. But I have prayed to you, and I have asked for your guidance, and here we are. How could something so sweet and true be so wrong, so misguided?

I put a housecoat over my nightgown, walked to Ottie's room, knocked softly, and looked around to make sure no one had noticed me, especially my cousin and her husband. When he opened the door and let me in, I felt like I was finally home.

We sat in bed for hours, clothes on, talking, embracing, holding hands, kissing, experiencing the aura of two beings as one. Then I took my head covering off and let my hair down—something Amish women reserve only for their husbands.

Ottie asked if he could brush my hair, I consented, and a spectrum of passion I had never felt before enveloped me. I could see the love and affection in Ottie's eyes as he ran the bristles through my hair. I could feel him gently caressing me with his other hand and lightly pressing his lips against the nape of my neck. I could hear him telling me he loved me and that he'd protect me.

At twenty-two I felt safe and secure, special and sublime.

And later, sometime in the middle of the night, when we'd lost track of time, the brush fell to the floor and we made love. Naturally. Unrehearsed. Unrushed. Without fanfare.

In the afterglow, a warm feeling of contentment washed over me, and we held each other for the rest of the night. Awake.

Neither one of us wanted the dawn to come.

Ten

I'd like to sum things up this way. What you did, we feel, was way wrong. But what Dad does to Mom or they do toward each other—and have done for years—is way wrong, too. Their own mistakes are making it hard for everyone around them. My constant plea and prayer is that we can all see our mistakes and truly repent.

—LETTER FROM WILBUR MILLER (BROTHER)

Freedom came to me June 8, 1996, a Saturday, ten days before Theodore Kaczynski was indicted by a California grand jury in the Unabomber case.

At about 9:00 that morning, Ottie gathered up Bertha and me and drove us to his house, where we retrieved some of his paperwork—research for a national directory he was compiling that listed people who drove for the Amish.

Because my father had prohibited Bertha and me from working at Ottie's house, the alphabetizing of the directory would have to be done at the farm. My father had also issued a warning: He wanted to talk to Ottie that evening, presumably about severing ties with him as the family's driver and prohibiting further foot treatments.

"What's up with your father?" Ottie asked after we'd arrived at his house and I'd told him about the warning. "What does he want to see me about?"

"He keeps ranting and raving that your divorce was in the paper and that people are talking," I said.

"Well, what's he gonna do? Stop you guys from working for me altogether? Stop me from driving for the family?"

"I think that's what he's up to."

"This is beginning to look like they're trying to get me out of the settlement," he said. "And you know, Irene, if that happens, I won't be able to see you anymore."

"I know, but I don't know what to do."

In the hollow space between us, sitting there on the couch in his house, there was the paralyzing realization that we might never make love again, might never share each other's company again, might never see our dreams through.

"There's only one thing to do, Irene," Ottie blurted suddenly. "We need to leave today. We'll leave, we'll get married in the Smoky Mountains of Tennessee, we'll settle down near my family in Kentucky."

"Can't we do this in a couple of weeks?" I protested. "Can't we have time to plan?"

"I don't think so, Irene. Now is the time."

"I . . . I . . . can't," I struggled, searching for another solution. "You know I can't leave the Amish. You know I can't leave Mom, and you know I don't want to hurt the rest of the family."

But no sooner had I said it that I knew instinctively Ottie was probably right. The timing was good; both of my parents had gone to town and would not be at the farm to interfere. Waiting, meanwhile, would run the risk that we'd be caught, either by our own ineptitude, circumstance, or my older sister speaking up.

I trusted Bertha implicitly, but she was the weaker of the two of us. She had been bad-mouthed so much by my father and others in the community that she had lost what little self-esteem she might have once had. It wouldn't take much to twist her arm into talking.

Another thought also occurred to me: Maybe my leaving would finally wake up the Amish community to the troubles at our house. Maybe, in an odd way, my leaving would actually help my mother.

Even so, I couldn't bring myself to tell Ottie I was ready to go. The desire notwithstanding, the words simply wouldn't form.

We packed up his documents, returned to the farm, and began unloading them with Benedict's help. Out of my brother's sight, I took a few dresses from my closet and put them in the van. Just in case, I thought.

Earlier, when Ottie had come to pick us up, I had also boxed up my crystal swans and put them in the van. I had begun to worry they would be discovered and figured Ottie could hide them better than I could. Subconsciously, the reason might have been far weightier.

When we were done unloading Ottie's papers, he suggested we go for a drive to continue our discussion. And so we did. Ottie up front in the driver's seat; Bertha and me in the back, as always, trying not to draw undue attention to a single English man in the company of two single Amish women.

My sister said little during the drive, occasionally interjecting mild protests in her soft, insecure way.

Once, she raised the issue of adultery. Ottie's divorces. His ex-wives still being alive and such.

"Is this really right?" she said.

"I'm tired of everything, Bertha," I told her. "I can't take it anymore."

Another time she said, "Don't you ever take your head covering off."

But I didn't say anything. I didn't tell her I already had. I just looked straight ahead until I came face to face with Ottie in the rearview mirror. He shrugged his shoulders, as if to say, "What will it be?" And I nodded.

I couldn't say, "Yes, I will marry you," or "Yes, I will go with you." But I'd managed a nod, and Ottie knew what it meant.

On the way back to the farm, I jotted a note to my mother on a scrap of paper. Two paragraphs, maybe three. Something to the effect of, "I'm leaving with Ottie; it's not your fault; you've been a good mother to me; I love you; Irene."

I knew that even if an opportunity had presented itself, I wouldn't have been able to tell her in person. The anguish creasing her face would have been too much for me to bear. At the same time, I felt obliged to let her know what I was doing.

We let Bertha off in front of the mailbox, I handed her the note, and we exchanged unceremonious Amish farewells.

"Goodbye," I said.

"Goodbye," she echoed.

She began crossing the dirt road to the farm. As she did, my father, who had returned from town, appeared from inside the barn, some sixty yards away. He looked at Bertha, then at the van. When he saw I was not getting out, he began walking toward us.

We didn't let any grass grow beneath the wheels.

Ottie headed down Gable Avenue, turned right on Johnson-Washington County Road, then made another right north onto Highway 1. Up the hill, out of sight, into the unknown.

Worried that we might be followed—that the police might be summoned by the Amish—we took the back roads out of Kalona. We could have taken the traditional route. Iowa City, then Interstate 80. Instead, we circled south and caught Highway 61 into Missouri and Illinois, bound for Glasgow, Kentucky, home to many generations of Ottie's family.

"Irene, honey, you'd better look back," Ottie said somberly, "because it could be a long time before you see the farm again."

"I know," I said, deciding against a last glance. "Please keep going."

I didn't want to subject myself to any more emotion than I was already feeling. Besides, it's hard for a person to move forward when they're looking backward.

Later, Ottie tried to lighten the mood when he reminded me I'd be able to assemble a new wardrobe.

"You can go shopping and buy anything you want. Satin and silk, frills and pastels. All the things you like."

"Yes," I said, smiling.

"And we can travel, and be alone, and one day, perhaps, raise a family."

"I know," I said.

"But if you're gonna be my wife, you're gonna have to do one thing. You're gonna have to shave your legs. You're not Amish anymore, you know."

"Right here? Right now?"

"Yep," he said, and he pointed to an electric razor he had brought along, one of the few possessions he had extracted from his house before we left. "You can use that if you like."

"You're sure?"

"I don't see why not."

Then he handed it to me, this English gadget of modern convenience, and I rolled down my leggings and began shaving the thin strands of hair from my calves.

I was nervous, just a little bit scared, and missing my family, despite our differences.

At the same time, Kalona all of a sudden seemed a long ways off.

Eleven

The only way I could come to you is if you come home with us. I don't think I could take it to see the one who inflicted so much pain and grief and made you commit adultery and to live in sin in the prime of youth!

—Letter from Mom

We arrived in Glasgow on Sunday, twenty-four hours after leaving Kalona. I had spent the journey in a peculiar state of tired anxiety.

Happy as I was to be on the outside, I was still being guided by Amish instincts, and I worried about the damage I had done to my reputation. Even though I knew I would not be returning to Kalona—at least not as an Amish person. Even though I knew there was nothing I could do to erase the tarnish.

Also hanging heavy in the back of my mind were the plaintive messages left on Ottie's phone in Kalona. We had stopped overnight in Mt. Vernon, Illinois, and Ottie had checked his voice mail. My sister had called. A brother, too. Even several bishops. All of them on the precipice of panic, begging me to come back.

They knew what was at stake, and so did I. If I married Ottie, an adulterer in their eyes, then later chose to return to the Amish, I would never be able to marry again. I would become an old maid.

And on that score, I had changed my tune. I no longer wanted to be an old maid in lieu of taking a chance on marrying the wrong person. I was sure I had found the right one.

En route to Glasgow, Ottie and I discussed getting married quickly—for two reasons. The first is that it might discourage my family and their friends from coming after me. The second is that making our union official—and moral—was paramount if we were to be held in God's good graces.

The contradiction of the latter point was more than apparent to me. We'd already made love, after all.

But I rationalized, as people often do. The night we gave ourselves to each other, I told myself, we did so because we thought it was the only opportunity we would have to share the ultimate human bond. Ottie would eventually leave, I would stay, and we would at least be left with the memory—and the comfort—of knowing we'd consummated our love.

Living together out of wedlock, on the other hand, was different. I had committed to Ottie and he to me, and there was only one choice.

We told Ottie's father, Ottie Sr., and sister, Faye, about our plans when we got to Glasgow. His dad, a spare man in his seventies with an iron will that had allowed him to survive cancer and a heart bypass, was to have surgery Monday morning in Bowling Green for a benign stomach growth. We would visit him in the hospital that morning, then head for Tennessee with Faye's daughter, Angela, and her husband, Chris, a professional wrestler in southern circuits. They would stand up for us at our wedding.

We would get our marriage license in Sevierville—between Knoxville and Dolly Parton's Pigeon Forge—then slip into the Smokies and get married in a mountain chapel. All well and good, we thought. But something happened along the way.

Ottie called Faye in Glasgow after we'd gotten our license and learned some Amish had called Ottie's most recent ex-wife in Indiana. My parent's plan was to get a group together, travel to the Smokies, and try to convince me to return. How they knew where we were going is a matter of conjecture, but they were aware the Smoky Mountains was a favorite of ours.

Everything changed in the wink of an eye. Ottie told Faye to make arrangements for us to marry Tuesday morning in a little white, nondenominational chapel near Nashville's Grand Ole Opry. Faye and her husband would meet us there.

None of this was how I had pictured it. In those rare moments growing up when I allowed myself to consider marriage, I thought of a large Amish wedding, usually on a Tuesday or Thursday. There would be a three-hour church service before the fifteen-minute ceremony. Three to four hundred people—including my mother—in the audience. A big dinner in the afternoon and socializing with the visitors, some of them from hundreds of miles away. And gifts for the bride and groom—tools and tack for him; towels, linens, and cookware for her.

Our marriage in Nashville took all of twenty minutes. There were six of my new family members in attendance: Ottie and myself, Faye and her husband, Angela (Angel for short) and her husband. The minister gave a brief sermon, we said our vows, kissed at the minister's command, and walked out.

We didn't have rings. Instead, we lit three white candles at the altar—one for Ottie, one for me, and one for our union. We didn't even wear our wedding best. Ottie had on casual slacks and a navy blue short-sleeved shirt. I wore my head covering, leggings, and a baby blue dress I had made for our Florida trip. The dress was fancy by Amish standards, but conservative by anyone else's.

Even in our free state, we were like fugitives on the run, making do with the final two thousand dollars Ottie had withdrawn from his bank, a stack of 1997 Amish calendars he had produced, the clothes on our backs (and a few to spare), the crystal swans in the van, God, and a yet unrealized dream of making a beautiful life together.

And one more thing. The love and support of Ottie's family.

Since our arrival in Glasgow, they had given me numerous chances to back out. They knew of my fears and wanted me to know that if I desired to return to the Amish, they would take me back. They didn't want me to feel trapped.

They showed their support in other ways, too. When we walked in the door at Ottie's father's house, the two men gave each other a bear hug. Then his father asked if he could give me one.

Silently, I thought, I don't know how to do this. But I reached out tentatively, and he put his arms around me. Welcome home, Irene. Welcome home.

The first place we headed after our wedding was Sugarcreek, Ohio. With all of Ottie's driving jobs gone, he wanted to pick up copies of a book he had compiled, *Amish Communities Across America*. The plan was to take ten thousand of them to Lancaster, Pennsylvania, the heart of Amish country, and sell them.

Fortunately, we called before we got to Sugarcreek. The proprietor of the print shop was liberal New Order Amish, but Amish nonetheless. He therefore felt beholden to the wishes of the Amish community at large, and someone had gotten to him before we had.

"Hello," Ottie said cheerfully after placing the call. "This is Ottie Garrett."

"Yes," said the proprietor, sounding strangely curt for a longtime friend.

"Well, you got my books ready?" Ottie said.

"No."

"Is there something wrong?"

"No."

"Well, when are they going to be ready?"

"I don't know."

"Then I guess there's no reason for me to be coming down there, is there?"

"I guess not."

Instinctively, Ottie knew something was wrong, said goodbye, and hung up.

We learned later that my family had driven to Sugarcreek and was listening on the other end of the line. I'm certain they had planned to ambush us at the print shop—where they would have an Amish audience—and create a scene of wailing, praying, kneeling, and raising their arms to shame me into coming home.

A friend later told us: "You sure dodged a bullet."

But truthfully, we hadn't. Within two weeks, the print shop owner told Ottie the Amish would never buy another of his calendars now that they knew what he had done. The owner threatened to participate in the boycott unless Ottie sold him the business for ten thousand dollars. It was a considerable loss, but Ottie had no other option. He was broke.

Ottie was also forced to sell the rights to *Amish Communities Across America* to a silent partner for fifteen thousand dollars for the same reason. A boycott.

We were newlyweds and we were losing everything. Our livelihood, my family—and soon our van.

Upon returning to Glasgow from our whirlwind wedding/brush with intervention, Ottie parked the van under a tree and left a trailer wire on the ground. Lightning hit the tree in the middle of the night, snaked to the wire, and fried the van.

Months later, when my family found out about our misfortune with the van, they said God was trying to tell us something.

"Yeah," Ottie cracked dryly. "Don't park under a tree during a lightning storm."

Twelve

Since you left, our family has reunited in togetherness and there is nothing between mom and I anymore! We are very, very sorry if we were the cause for you to leave us. I ask you to forgive me in all areas where you feel I have wronged you or mom.

—LETTER FROM DAD

The letters began arriving shortly after we moved into a tiny, $200-a-month square of a house at a crossing in Uno, Kentucky, northeast of Horse Cave.

They would become a constant reminder of where I had come from and what I had done. They would—and still do—tug at my conscience, testing my will and my conviction.

The first was dated June 14, 1996, and it included remarks from every member of my family. My father began the missive and the rest followed:

To Ottie and our dear dear Irene:

Greeting in Jesus' name.

We with broken hearts want to tell you we are very sorry for anything that would have caused you to leave our home, Irene.

Since you left, our family has reunited in togetherness and there is nothing between mom and I anymore! We are very, very sorry if we were the cause for you to leave us. I ask you to forgive me in all areas where you feel I have wronged you or mom.

Irene, you are always welcome in our home, and come before it is too late! Never think the Lord cannot forgive the sin of adultery.

There is forgiveness where there is true repentance!!!

Pray for your salvation and we and our church are all praying for you.

Your broken-hearted father,
Dad

My mother was next.

From your heart-broken mother:

Irene, this is the hardest thing in life I've had to face. Please, please forgive me for anything I've done or said that caused you to do this. Dad and I have made amends and the family, but it's so lonely, dark, and empty without you. I don't see how we can face people anymore.

The neighbors, relatives, and friends have come to comfort us. . . . It was very touching.

After church, the people didn't eat much and were very quiet. At the singing, they sang mostly wake songs. Everything is so sad.

O Irene, you can still come back. People are all praying that you will. People are so concerned about you. Can't you feel the prayers and the tears for you? . . . I feel so torn up, weak, and sad. I also have heart pains, so I don't know if I'll get to see you here again, but I will never cease praying for you as long as life permits. . . .

Then my sister.

My dear and only sister Irene:

It is so lonely here without you to go to bed with. . . . I'm sorry for all I did wrong. The song has left our house. The nieces and nephews don't know what to do without you!

Please forgive me if I did anything wrong.

Your only broken-hearted sister,
Bertha

And my brothers.

Things are different, Irene. You can't imagine it.

We hope we can all work together from now on and we hope and pray that someday you can be here to help us. We miss you!

Sincerely,
Wilbur

Dear Irene:

O Irene, please come home!

I just about can't put it in words how I feel. Please, oh please, forgive me for what I have done wrong. Please give us another try before you go too far!

You can't imagine how the days were spent Sat. eve, Sun., Mon., and Tues. Lots and lots of tears and many, many friends came to share their sympathy on Sun. and Mon.

O please come home, Irene!

From your brother full of faults,
Benedict

Dear Irene,

If words could only write the deep grief of my parents, how I would try to put on paper what they are going through. And I could find forgiveness if I had a part in you doing this. . . .

Please don't feel anytime that you can't face the ones at home because you are welcome.

Your brother wishing you were home,
Aaron

Dear Irene:

O please come home. Our home is so empty. We miss you very much.

From your brother,
Earl

O please Irene,

Please come home and try this reunited family. You cannot imagine the difference.

Hope to see you soon!!! We still love you greatly.

Elson

In all honesty, I couldn't imagine the strife within my family had vanished between the day we left (June 8) and the postmark on the letter (June 14). It seemed inconceivable that my parents could patch up their differences in six days, or that my brothers and sister could forget all that had gone on before.

In time, I would discover my hunch was right. I learned that my father would preside over letters sent to me, dictating the content and context of the messages. Any appearance of unity, therefore, was rehearsed.

Most of the letters were also addressed to "Ruth Irene Miller," or "Irene Miller," or "Irene, c/o Ottie Garrett." The latter was rare. More often than not, it was clear my family did not want to recognize Ottie's place in my life, and perhaps even desired to drive a wedge between us.

Worse, I learned, when I'd send birthday and holiday cards to family members signed Irene and Ottie, the correspondence would often be burned.

Then there were the letters that can only be described as cruel, particularly one from my father after I'd tried to explain why our marriage was sound in God's eyes:

> *We received your letter yesterday. . . . It broke our hearts again to see you try and use scripture to cover such an evil deed. Mom just cried and cried and finally she said if only you could have died when you fell out of the upstairs window. You wrote you have not died. We hope and hope you can repent before you're spiritually without life.*

The reference to the fall cut through me like a scythe.

I knew my mother couldn't possibly have said it. And if she had, it almost certainly was at the prompting of my dad.

In any event, to tell a child you wished they had died—or wished they'd never been born or wished you'd never had them—has to be one of the harshest forms of mental abuse.

I could try to put an Amish spin on it and say the remark was similar to the one my father had made about my brother after little Tobias died. Tobias would never again be tempted by the devil. I would never again be tempted by the devil.

But I took it differently. I had so disgraced them by leaving the Amish that my death was preferable to my life.

I was about Tobias's age, three or four, when I fell from the window. Summers, Bertha and I would move our bed next to the window so we could benefit from the nighttime breezes. One night, in my sleep, I rolled over against the screen and plunged two stories, headfirst, into a cement well at the edge of a flower bed. The force of the impact knocked me out.

When I came to, I walked around the house to the front porch, knocked on the door, and was let in by my parents. I told them what I thought had happened, but we didn't seek immediate medical attention beyond visiting a chiropractor.

About a year later, I began having horrible headaches and stomach discomfort, and my parents took me to a hospital for tests, which proved inconclusive. When the symptoms continued, we consulted an Amish doctor and he concluded I had fractured my skull. The bone on one side of my head, he said, had developed a ridge at the break.

He put his hands firmly on my head and began pushing, trying to re-break the bone so it could set properly. The pain was excruciating and the recovery long. It would be several years before the headaches stopped.

My parents' somewhat guarded attentiveness to my injury was common; the Amish would rather nature take its course than intervene. My mom broke her arm once and when the doctor told her to come back for checkups, my dad said, "Can't you just let nature tell her if she has to come back or not?"

Part of it was the reliance on nature, part of it was he didn't want the medical bills to run up. In the end, the doctor gave in. My mother didn't go back.

Sometimes, my father would seem to soften in his letters, although they always sounded as if he was more concerned with himself than with others: "I would seriously long to see heaven when my life is over and I need peace with you," he wrote.

Sometimes, he'd even tell me he loved me—something he'd never done when I was at home.

But I couldn't escape the years of abuse or the years of suggesting a problem was someone else's fault, not his.

If he could lay the blame elsewhere, it somehow absolved him of any wrongdoing. This was evident when Rick Farrant, my coauthor, visited his farm in the spring of 2000 to talk about the book.

My father opened the front door with a handshake and wide smile that turned to an icy stare when he learned the purpose of the visit.

"I don't want any part of it," he said. "I just wish you'd cancel the whole deal. That would be a blessing to me."

"Why do you want the book canceled?" Rick asked.

"Because it's wrong. It's evil."

"But it's important to get both sides of a story. And you have a side."

"I've said this before. I don't want any part of it."

He looked down and shook his head several times, a mannerism he would repeat often during the next fifty minutes of tortured conversation.

Then, as he leaned against the partitioned entrance to the living room, thumb and index finger fastened around a suspender strap, he asked if the discussion was being tape-recorded. It wasn't at that moment. And now, it wouldn't be. The record would be left to the author's years of practiced recall; the kind that is jotted down the moment an interview is over.

Alvin talked about how he had tried to be a good father while at the same time upholding the strict rules of the Amish church.

"The scripture," he said, "doesn't bend to me. I'd like to think I bend to the scripture."

He denied being abusive to his children, said that "of course" he loved them, and suggested that if Ottie hadn't come along, I wouldn't have left.

"If Ottie doesn't come here," he said, "I have to believe Irene's still here."

Asked if I might not have played some role—of my own volition—in leaving the Amish, he turned to my mother, who was sitting at the dining table, elbows planted firmly on its oak top, head in hands, face covered.

"Well, what do you think, Mom?" he asked her.

That was another thing he would do over and over during the conversation, as if he were trying to shift the tough questions—and perhaps some of the blame—to her.

Sometimes she would answer. More often, she would remain still, leaving a painful silence in the room.

"Can you think of any reason why there's such a rift between you and Irene?" he was asked.

"No," he answered. "Mom, can you?"

No response.

Later, referring to the news articles, book, and perhaps a movie that would chronicle my life, he said, "There are a lot of people who've left the Amish, and they don't do this."

"Does that mean your reputation among the Amish has been tarnished?"

"No," he answered. "Mom, can you think of any way it has?"

"If it has," she said, "we haven't heard about it."

How strange they would say that, I thought later. In one of his letters, my father had written in German:

> *It hurts so much that I have a daughter that is disobedient! And the New Testament says how the minister's children should be obedient, and what a mark you make on us. And how serious when I am supposed to stand up in church again! The 'mark' that we and our children carry now hurts!*

At the far end of the dining table, a pair of eyeglasses sat atop an opened Bible, keeping the reader's place. A yellow and green parakeet chirped busily in a cage hanging from the ceiling in the modestly appointed living room. Bertha, doing chores, periodically entered the spotless kitchen off the dining room to listen, but dallied only briefly before shuffling away. She smiled once at the visitors. My parents, backs turned, couldn't see her.

"Do you still want Irene to come back?"

"Yes, I want her to come back," my father answered. "How can she think I don't?"

"But if she comes back, she can never marry again. Isn't that right?"

"If she can humble herself to come back, God will take care of that," he said. "She will lose her taste for marriage."

"She will?"

"She will."

But if there was anything I would lose my taste for, it would be for the sometimes inflexible, narrow-minded ways of the Amish.

Thirteen

I feel so sorry for your family! I wish you could get a glimpse of the great sorrow and grief they are suffering!! Your mother is going downhill and Dad looks pale-faced! Oh how sad! I hope and pray that they will not lose their minds through all this!

—Letter from Perry Miller (Uncle)

They say a person can't truly know something if they haven't experienced it, or can't acquire perspective if they haven't been on the outside looking in.

Both of these notions would serve me well as I entered the English world and saw for myself that not all English are evil—and not all Amish are good.

Ottie said I was like a sponge, soaking up everything I could about my new life as fast as I could.

"Honey," he would say, "slow down. It ain't gonna go nowhere. You've got time to learn."

But I couldn't be swayed. I had prayed to God, asking that he allow me to keep an open mind about what I would witness on the outside. Confident in his guidance, I became a magnet, drawn to every little detail, thirsting for a knowledge denied for too many years.

I marveled at how friendly everyone was. From Ottie's family, to strangers on the street, to the minister and congregation at our new place of worship—Holy Trinity Lutheran Church in Bowling Green, Kentucky.

The church was truly something to behold. Families sat together, creating a warmth of spirit absent in Amish services. When the Amish gather for church, the men sit with the men, the women with the women. It is a segregation born of centuries of tradition.

I was also amazed when, at the close of services at the Lutheran church, the Rev. James A. Bettermann hugged every parishioner who walked the greeting line. Such affection and caring I had never seen before in God's house.

Nor had I witnessed another grand religious ceremony of the South. Some Saturday nights between 7:00 and 10:00 P.M., people would cart lawn chairs to a department store parking lot in Glasgow and settle in for a night of listening to free gospel music performed by local musicians and singers. Marvelous, I thought. Simply marvelous.

I was equally impressed one day when Ottie pulled to the side of the road as a stream of cars with their lights on passed in the other lane.

"Why are you stopping?" I asked.

"Because it's a funeral procession. That's what we do to honor the dead."

Honor. What a nice word, I thought.

Among the Old Order Amish, such reverence would be unheard of. They believe that when a person dies, there is no sense in paying extended homage to them because they aren't there to witness it. They don't even place flowers on graves.

Beyond the compassionate social customs of my new

world, I also reveled in the modern conveniences and entertainment offerings. The electric appliances. The cars. The movies. Television.

I instantly fell in love with washers and dryers, which could not only clean clothes spotless but do it quickly. Back on the farm, we'd have to fire up the gas generator to pump water into the washing machine. And when we dried the clothes, it was on an outdoor line. Washing clothes was such a chore, we usually did it only once a week.

My first movie in a theater, *Dragonheart,* starring Dennis Quaid and Sean Connery, was a bit unsettling, both because we were sitting together in the dark and because the violence of the medieval fantasy was so graphic. I winced with every slash of a sword, every spurt of blood.

In time, though, I grew to enjoy watching movies in theaters—and at home on television. Especially the old Westerns, and especially those featuring John Wayne. They seemed so true, so American, and so full of the right values.

Television and the 1996 Olympics also offered my first chance to hear the National Anthem, and I was at once drawn to its beauty and power, experiencing for the first time a pride in America. The Amish don't play the National Anthem. They don't recite the Pledge of Allegiance, either, feeling perhaps that both overshadow God's importance.

A seemingly more mundane matter—shopping for clothes—took a little more getting used to. I had Faye to help me, but my naïveté overwhelmed her at times. Because we had made all of our clothes on the farm, I was completely oblivious to clothes-buying etiquette when I stepped into a department store for the first time. And the second time, and perhaps the third.

The wonder of it all was that there were so many patterns and colors to choose from. The frustrating thing for Faye was that, despite the broad selection, I initially gravitated to clothes almost as conservative as my Amish attire.

The first time out—I think it was a Sears or a Penney's—I bought a navy blue dress with shiny, satin-like material and a rather plain peach dress. The hems on them weren't much higher than my ankles.

I also got a pair of white tennis shoes and white socks—white, because I was so tired of wearing black.

I didn't try the dresses on to see if they would fit because I felt uncomfortable with the idea of doing so in public. I didn't know about dressing rooms.

Faye, a stout, friendly blonde with a cackle that fills a room, would say, "Go ahead, try it on."

And I would hold the dress up to me and say, "No, it looks like it will fit me."

When I finally got the nerve to try a dress on—several shopping trips later—the results were embarrassing. While I was looking at one particular rack of clothes, Faye turned her attention elsewhere. When she looked back, I had put a new dress over the clothes I had worn into the store.

"No, no, honey," she said, struggling to hold back laughter, "we don't do that."

"But if it fits over my clothes, it should fit me just fine," I protested innocently.

"Sweetheart, we have dressing rooms for that."

Faye would later say she had to leave the store briefly, go to the van, and compose herself. I guess Ottie and Faye had a good chuckle.

Another time, having mastered dressing rooms but still short on protocol, I wore a new dress to the checkout line and paid for it, showing a surprised cashier the tags hanging from the sleeves.

Fortunately, things got better over time. I graduated from modest clothing to attire with more pizzazz, although once I'd found a pattern I liked, I wanted to buy it in all the colors it came in. Kind of like a man buying dress shirts or slacks. One pattern, one maker, every color.

I also began buying touches of makeup, perfume, and facial cream, things I had avoided early on because I thought such aesthetics silly and a waste of money.

"Facial cream," Faye said in her sultry Southern drawl, "helps stall the aging process."

"Maybe Ottie doesn't want me to have it," I quipped. "Maybe he'll want me to age."

"I don't think so, honey," she said.

She was right, of course. Ottie does like a little dab of feminine allure. Even in clothes, I learned.

Shortly before Christmas, six months removed from the Amish, Faye suggested on a shopping trip to Nashville that we stop at Victoria's Secret. Ottie, she said, would be most appreciative if I bought some sexy lingerie as an early Christmas present. And so I did. Nervously, but knowing full well he would be the only one to see it on me.

That night, after we got home, I went into the bathroom, put on the red-and-white lace teddy, black garters, white gloves, and red feather boa I had bought and, feeling more than a little silly, strode awkwardly to the side of our bed.

I could tell by the smile on Ottie's face that he was pleased, that I looked rather fetching.

"That's a memory I'll take with me, that's for sure," he would say later, grinning mischievously.

Fourteen

> *Today is Loretta's birthday. We are planning to freeze ice cream for supper tonight with the fresh snow. . . . In a couple of weeks we will have our ninth wedding anniversary. (Time goes on) but we don't know how long. Ann will be 2 in September. She is more fine-featured than the rest were. Her hands are no bigger now than Harold's were at birth.*
>
> —Letter from Elson (Brother)

Of all the letters I received from my family after leaving, my oldest brother's were the most consistently kind. They were the kind of correspondence you'd expect from a relative, filled with warm updates about a family's day-to-day activities and rarely carrying a judgmental air. On one of his letters, he even had his children trace their hands so I could see how big they'd grown. They wrote their names in the white spaces of the palms and each drew a little smiley face.

I've heard Elson took flak for choosing to be nice—to be brotherly in his letters—especially from my father. But Elson, bless his heart, is his own person, determined to right the wrongs of previous generations.

On any given day, one can find Elson working in his cozy, odorous blacksmith shop behind his house along 110th Street, holding court with the English who come by word of mouth to have their horses shod. There, he melts globs of carbide onto glowing orange, hot-fired horseshoes—drill-teching, they call it—to reduce the wear on the shoes, and on this day he is refitting a huge, tan Belgian draft horse that is cinched tightly in a harness.

He gets nine dollars a shoe for the drill-teching; twenty-eight dollars a shoe for the refitting, which involves digging out, clipping and filing the hooves, shaping the shoes with a hammer, and deftly driving nails into shoe and hoof.

Elson's work has given him the forearms and biceps of a professional wrestler and, with his reddish beard, sunglasses, hat, and bulk, he looks a little like country-and-western singer Hank Williams Jr.

But he is neither fighter nor reveler. He is a dedicated thirty-something husband who recognizes—and in some cases deplores—the contradictions and biases of the Amish faith.

In that way, we are alike.

He has simply chosen a different path. His calling from God, he says, is to stay among the Amish to help them with their problems, and to forgive our father for his transgressions, something I continue to struggle with.

Elson has a sense of humor that is rare among the Amish, and he sometimes used it to brighten our childhoods. One summer, when I was ten, Bertha, Aaron, Elson, and I were walking a gravel road to deliver a cake to an elderly Amish woman when Elson suddenly made a beat toward an electric cattle fence.

"You suppose it's hot?" he asked.

We all nodded.

"Bet it's not," he said, and he put his hand on the fence to prove it.

"Well," he announced, "it didn't shock me, so it must not be hot."

He asked us if we'd like to touch it, too, and Bertha, cake in hand, reluctantly agreed after declining several times. But when she put her hand on the fence, a surge of electricity coursed through her arm, she let out a squeal, jumped several inches off the ground, and dropped the cake.

Elson had fooled her, purposely failing to show how to time the syncopated pulse of the current by listening to it.

Even today, Elson possesses a playful wit. His job, he'll tell you, laughing, takes a strong back and a weak mind, the latter referring to the apparent stupidity of embracing an occupation that requires so much labor.

Still, he loves his public work. And he loves his private life as an Amish husband to Loretta and father to six children.

He dusts off the seat of a white plastic chair near the door of his shop and begins talking about growing up in the Miller family. A rooster crows brazenly just around the corner. Skipper, a scrawny, cat-sized, black-and-white dog with unbridled energy, scampers in and out of the shop, collecting white hoof shards and using paws and snout to bury them in the mud outside.

A doctor from Iowa City used to visit to buy the shards at five dollars a bag so he could feed them to his dogs—in lieu of store-bought milk bones.

"I don't really know if they have any nutritional value," Elson says, shrugging his shoulders.

When he talks, Elson is deliberate, trying to make every word count, every thought both humble and accurate.

He'll tell you he was beaten by our father with leather straps when he was a child, usually for accidents that couldn't be helped. Like breaking a window. They weren't well-thought-out whippings to make a point, he'll say. They were an inflicted pain born of blind rage.

"I know Irene had her struggles with Dad, but really, plain to speak, I don't think she knows what it was like to be abused like I was."

Our father, Elson will say, had little tolerance for people's weaknesses and would often poke fun at people—his children included.

To this day, he'll tell you, he has trouble controlling his own rage, and on several occasions he's had to apologize to his children when he's flown off the handle. He prefers to "have a listening ear and to work things out together." But rage, he says, is still his first impulse, even though it's something he doesn't want to repeat.

"It takes more than just deciding not to be that way. It takes a lifetime, I think. I'm not expecting that it's just going to fade or go away on its own.

"The biggest thing is if you believe in the grace of God, I guess. That's the only way I can handle it. God truly helped me to forgive my dad.

"And I think, you know, as far as for me, logically speaking, I think Irene's biggest problem was she couldn't stand it. And as far as for her, she can't forgive dad for what he did.

That must be what she's still working on evidently, or she wouldn't. . . ."

His voice trails off but the implication is evident: I wouldn't still be away from the fold.

He says he believes that Ottie and I can make it as a couple. His only concerns are that I married a much older man, and one whose ex-wives are still alive. As understanding as Elson is, the latter still constitutes adultery in his world.

But on other matters of Amish doctrine, he is less understanding. He doesn't believe that only the Amish can be saved.

"There's more and more Amish people feeling that way," he says. "The more people I hear making remarks like that, the more it infuriates me. And for me, I got no problem telling people how I feel about that. There are some people that avoid me for that fact.

"I guess I feel that if they do have that attitude, they don't have any more chance than the outside people. God's gonna look up everything. And if people have that attitude that they're gonna have a chance above anybody else, they missed the point already. That's totally taking something into their hands that they have no right to."

Elson will also tell you that people who criticize the Amish for not doing missionary work outside the community have a good point, although he has an answer for the naysayers.

"If I want to be truly forgiving and be a Christian like Christ was, if I want to be Christlike, I have to lay things down . . . the biggest mission field is here at home, isn't it?

Don't we have many Amish people in need, just like they do on the outside?

"Just the very people who think they're the only ones who are saved are the ones in need."

A bell chimes several times outside the shop where Elson sits, startling a visitor.

"Is that the dinner bell?" the visitor asks. "Yeah," Elson says. "But that's just the children playing for the sound of it."

He goes on, without missing his place: "We're all born with the same chance. It's up to us if we're going to try to live for Christ, or we're going to try to live for ourselves and try to hurt other people."

Ten years ago, he says, he had a nightmare that had a profound effect on him. It was one of those dreams that makes a person suddenly sit upright in bed, trembling.

"I dreamed that the world had evaporated and I was lost. I was going to hell."

In the final analysis, he says, God will assess a person's character and commitment and decide who goes to heaven and who goes to hell.

"After all, God is the one that is going to right the end. R-I-G-H-T. He will right the end."

Fifteen

If you have questions about the ban and marriage, the answers are plain in the prayer book you have. It's in the articles in the prayer book. The last couple of pages are the list of ministers and elders who read, approved, and adopted these articles as scriptural.

—Letter from Mom

The letter I knew would come arrived about six weeks after leaving the farm. It was from my uncle, Elmer T. Miller, the bishop of our community, and it formally closed the door on my past.

Dear Irene:

Greetings in Jesus' Holy name, unto thee every knee shall bow and every tongue confess.

Had a nice shower today and cooled off nice. Oats are in shock and corn is growing, making a beautiful scene of God's creation.

It is now past six weeks that you left home, family, neighbors and church—in search of ???? This leaves a hollow, empty spot in the above mentioned places and also in my heart, as I at times think it can't be true.

According to the scriptures, we cannot feel this is pleasing in the eyes of the Lord, as Jesus himself spoke: 'That whosoever shall put away his wife, saving for the cause of fornication, causeth her to commit adultery, and whosoever shall marry her that is divorced committeth adultery.' Matt. 5:32. In Gal. 5, verse 19, we read of the works of the flesh, and adultery is named among them and concludes by saying that they that do such things shall not inherit the Kingdom of God.

We may think that God forgives, but God only forgives when we repent, and unless we repent we will never be able to inherit the Kingdom of God.

As you well know, it is the duty of the church not to tolerate such things in the church. So, by voice of the church, you have been excommunicated from the church. This was done yesterday.

You may now think that nobody likes [you], nobody wants (you), etc., but wait, my heart still aches for you and your soul as many a tear has been shed since you left.

Oh that you may repent while living in the days of grace.

This is written out of love and concern.

Your neighbor, uncle, and bishop,

Elmer T. Miller

The verdict was hardly a surprise. In fact, it was somewhat anticlimactic. The real gut-wrenching stuff had occurred several weeks earlier when my family had called our house in Uno to tell me precious little time remained before the church rendered its decision.

We had heard through others that the Kalona Amish were convinced I had been drugged and held against my will, or that I had become pregnant and fled out of embarrassment. It was logical, then, when my parents repeatedly asked me during the phone call if I was okay.

"Did he give you drugs?" my father asked.

"No."

"You sure he didn't give you a shot or something?" my mother inquired.

"No. Everything's fine."

Most of the conversation was not as lucid. My parents spent a lot of time crying and wailing, and that brought tears to my eyes and made it difficult for me to speak. I knew the only way I could ease their sorrow was to say I was returning, and that was not something I was prepared to do.

By the time Elmer got on the phone, I was so devastated, I could do little more than listen and sob as he discussed the church's stance on shunning. It is spelled out in the Mennonite Confession of Faith, the same document I had been baptized under:

"As regards the withdrawing from, or the shunning of, those who are expelled, we believe and confess, that if any one—whether it be through a wicked life or perverse doctrine—is so far fallen as to be separated from God, and consequently rebuked by, and expelled from, the church, he must also according to the doctrine of Christ and his apostles, be shunned and avoided by all the members of the church (particularly by those to whom his misdeeds are known), whether it be in eating or drinking, or other such like social matters. In short, that we are to have nothing to do with him; so that we may not become defiled by intercourse with him, and partakers of his sins; but that he may be made ashamed, be affected in his mind, convinced in his conscience, and thereby induced to amend his ways."

The ordeal continued when my brothers and sisters got on the phone. They tearfully described how my leaving had cast a pall over the family, how my parents were losing their minds and might have to enter a mental institution, how my mother was experiencing heart pains and didn't know how much longer she could hold out.

"She takes sleeping pills at night," one of them said, "energy pills in the daytime, and she doesn't eat anymore."

"What has Ottie done to you?" another asked.

"Nothing," I said, weeping. "I love him."

I stood up to the barrage only once—when my father recited a biblical phrase he had mentioned several times in his letters:

Ephesians 6:1–3: "Children, obey your parents in the Lord: for this is right. Honor thy father and mother; which is the first commandment with promise; that it may be well with thee, and thou mayest live long on the earth."

I told him bluntly he should read the next verse, too:

Ephesians 6:4: "And, ye fathers, provoke not your children to wrath: but bring them up in the nurture and admonition of the Lord."

From that moment on, he never again raised the issue of honoring mother and father. He must have known he couldn't pass muster with the fourth verse.

Later, I bristled at other inconsistencies in the behavior of my family and the rules of the church. My uncle had said in his letter that the ban was done "out of love and concern," but there is nothing loving, nothing Christian, about shunning.

Those who are put in the ban and remain living among the Amish are ignored in social settings. When meals are served, they eat separately, and others are instructed not to pass food to the accused, lest the sins should pass from hand to hand. The only acceptable excuse to talk with someone in the ban is to try to show them the error of their ways and encourage them to repent. In short, the life of an excommunicated person is a lonely one, akin to solitary confinement, or worse, a leper colony.

I once mentioned to my father how ironic it was that the Amish follow the teachings of Jacob Ammann, who himself was excommunicated, albeit voluntarily. But my father dismissed it as nonsense. Most Amish either don't know the history of Jacob Ammann or choose to ignore it.

I was also miffed that my parents had not tried to see me in Kentucky, even though we had given them our phone number and address, and invited them to visit. If they, in fact, believed I had been shanghaied, wouldn't they do everything in their power to free me? Wouldn't they hold off on the excommunication until they could be sure of my circumstances?

I don't know the standard length of time before shunning someone who's left, but I've been told by other Amish that six weeks is on the hurried side. Why the rush, especially when so much was at stake? Why didn't they come see for themselves first?

These thoughts further galvanized my determination to stick with my decision and make a life for myself on the outside.

And they softened the blow just a little when the bishop's letter arrived.

I was tired of the drama, disturbed by the contradictions, and reasonably certain that God would not abandon me in my time of greatest need.

The Amish believe that if a person dies in the ban, they will forever be condemned to hell. It is yet another fear tactic to keep people in line.

But it seemed to me that the God I knew wouldn't approve such punishment—that his own son, when he was on earth, never turned a person away. The God I knew was kind, considerate and forgiving. And I was praying to him still—at home and in church.

"Well, I guess it's official now," I said after opening the bishop's letter.

I had no tears or regrets—at least not on that day. It was over.

But only for the moment.

Even though I was no longer Amish, I felt like a cast-off, and there was a part of me that wanted to find a way to have the ban lifted, if for no other reason than to bring a symbolic peace of mind.

Just as disturbing, I didn't know when I would see my family again.

Sixteen

The choice you have made causes pain in my heart deeper than the loss of my companion!!! . . . Love not the world! Its dazzling show conceals a snare of death.

—LETTER FROM MRS. TOBIAS MILLER (GRANDMOTHER)

About a year after arriving in Kentucky, we moved out of our little rented house in Uno—the one with the bathroom tacked on when outhouses went out of style—and into a somewhat larger, $375-a-month rental along Old Dixie Road near Horse Cave, just off Interstate 65 in Hidden River Valley.

We call it the Home Place and it is in a pastoral little strip of hollow (holler) with sparsely situated, modest homes and gently sloping meadows. It is an area of language oddities where tourists come to visit the labyrinth of caves that wind underground, where tobacco (tabacca) fields and warehouses dominate the landscape and the economy, where driving a rock-bottomed creek (crick) to get somewhere is common, and where ferries carry cars across the Cumberland and Green rivers because it's apparently cheaper than building bridges.

A person can still buy illegal, gut-busting, 120-proof moonshine in the hills of otherwise dry Hart County if they know where to look, and the Confederacy—a remnant of a war the Amish detest so much—is still very much alive.

North of Horse Cave is the site of the Battle of Munfordville, where 50 died and 307 were wounded when Confederates attacked a Union fort on September 14, 1862. A year earlier, Brigadier General John Hunt Morgan and eighty-four of his followers were sworn into the Confederacy in Munfordville. The marauding band of fighting men went on to acquire a small measure of Civil War fame as Morgan's Raiders.

The little town along the banks of the avocado-tinted Green River is also home to the Old Munford Inn, where Andrew Jackson stayed in 1829 en route to his inauguration as the seventh president of the United States.

Close by is the site of the Battle of Rowlett's Station on December 17, 1861.

Kentucky was among the states beset by divided loyalties, where people from the same families often fought on different sides of the war. The sting of those divisions—and of the Confederate loss—are as acute as if it happened yesterday, much the way Tony Horwitz describes the vestiges of the conflict in *Confederates in the Attic.*

In a cemetery overlooking the deep hollows of the Munfordville battlefield, weathered gravestones in the shadow of a giant, knobby elm bear witness to the festering sentiments. Some are honored with American flags, some with Confederate. On one grave, both flags fly.

There is even evidence of the war several hundred yards from our home in Horse Cave. Cloaked in a grove of pine trees on the other side of Old Dixie Road is a monument to a fallen soldier of the Confederate States of America. The memorial is a composite of geodes and mortar with an American flag positioned at the top. It tells of an unknown warrior who fought under General Clay Anderson, Division 11th, Louisiana, and who perished September 9, 1862. It was erected in 1934 by one Sam Lively.

Still sadder is a monument erected south of Horse Cave near the Tennessee line in the parking lot of the Alpine Motel overlooking the Cumberland River Valley. In the lot's median—at the foot of a loosely piled bed of shale and a weathered gravestone—is a metal sign that reads:

"CAPTAIN JACK McCLAIN, COMPANY J, 1st KENTUCKY CAVALRY. Braver men never responded to boots and saddles than the 1st Kentucky Calvary. Jack McClain accidentally killed a good friend. In sorrow he took his own life September 21, 1866. Previously, McClain had requested, "When I die, I want to be buried on top of that highest hill overlooking Burkesville, as that is as near heaven as I will ever get."

Even nonbattle tragedies involving southerners are memorialized here in the name of lingering patriotism.

Northerners aren't exactly in danger in these parts, but they aren't exactly loved, either. People in Kentucky have a saying that there are two kinds of northerners: Yankees and damn Yankees. Yankees are the ones who come to Kentucky and keep on going. Damn Yankees are the ones who stay.

You get the picture.

Visitors should know, too, that nearly everyone possesses a gun of some sort. Saturday nights in nearby Cave City,

Kentuckians even auction guns—among other things—from a downtown storefront that promises "Bargains Galore" in red lettering on a street-side window.

For some people, the auction is the big event of the weekend, seemingly revered more for its social attributes than the lighting fixtures, plates, and jars of old marbles that pass the bidding table.

Ottie himself has a .40-caliber Ruger that he keeps for protection. A big, black, semiautomatic pistol with a kick that can easily jerk an unsteady pair of hands skyward and a power that is deadly.

One of the first things Ottie did upon our arrival in Kentucky was teach me how to fire it—in case I ever needed to fend off a mugger, robber, or burglar. And early on in my training, he pronounced me a natural. A sure shot if there ever was one.

It sounds odd for a pacifist ex-Amish who had never before held a gun in my hand. But then, I was eager to experience everything in my new world. Moreover, I had already bought into his argument about the need to maintain one's liberties, even if it meant shooting someone, even if it meant going to war.

I was also quick to obtain my driver's license, realizing that mobility was an integral part of the English world, if not just a wonderful, efficiently fast creature comfort. In fact, it seemed to me that driving an automobile was a darn sight safer than steering a buggy. At least there wasn't an unpredictable horse in front of you that might take off in the wrong direction at a moment's notice.

We had purchased a new, smaller van after the other one burned, and I practiced driving it in the Haywood Acres subdivision owned by Ottie's father on the outskirts of Glasgow. It is the same subdivision where Faye lives, where the streets are named after Ottie Sr.'s grandchildren. Angela Court. Dana Court. Phillip Drive.

I'd circle the subdivision over and over while Ottie and his father watched from the red-brick patio out front. For at least an hour at a stretch, or until I tired of the routine. Whichever came first.

Six months later, I got my learner's permit, and six months after that my formal license. I was a natural at driving, too, although somewhere along the way I acquired a lead foot. Just like Ottie.

It was a land of discovery, this new place of mine. And a safe one to explore from our little den along a narrow stretch of blacktop that once served as the main highway—our white house with the four clipped soft maples in front, the porch on the side with two hummingbird feeders and a wooden sign above the door that says "Ottie & Irene Garrett," and the assemblage of feeders along a fence that attracts goldfinches, cardinals, bluebirds, orioles, and sparrows by the dozens.

On the other side of the fence is my second sanctuary. My garden. It is an ambitious 50-by-75-foot plot tilled in the rich, red Kentucky earth, and each year it bears lettuce and radishes, peas and beans, corn and zucchini, tomatoes and pumpkins, all of them bordered by bright yellow marigolds and deep-red cockscomb.

It is here that I find solitude and an unmatched closeness with nature and God. Sometimes, Ottie goes out on the porch, props up his bad leg, and talks to me while I weed and harvest. More often, though, I am alone. With my thoughts. And prayers.

There is a perfect symmetry to the things in the garden, the reaping and sowing, the spring rains, the plants growing. If there can be one place where God has put his hand, it must be here, because in the garden, there is no turmoil or conflict. There is only hope, promise and, if all goes well, the sustenance of food at season's end.

The fresh fruits and vegetables are our main staples in the summer and fall.

Winters, we eat what I have canned.

Some people might assume my garden is abundant because I was once Amish and therefore a prodigious farmer. But truth be told, I have discovered a wonder of the modern world that helps the plants become robust.

Miracle-Gro.

No self-respecting Amish person would ever use the stuff for gardens. But then, that was no longer a concern of mine. If something works, why not?

Beyond the health benefits—to soul and body—the garden helps save money. And that's also important. Especially when you're living in a one-income family, and that income is Ottie's disability pension—and whatever business projects we can develop.

In the first year out of Kalona, Ottie contracted with a local Amish artist to use Ottie's photographs in painting scenes for yet another Amish calendar. Only the artist's name—and a company moniker (Country Roads Art) in my name—appeared on the calendars, thus keeping Ottie's involvement a secret.

But even that enterprise went the way of the others. The artist eventually threatened Ottie with a boycott and Ottie had to sell the business at a loss, or lose everything.

We were back to square one.

Ottie would—and still does—bring in a little money playing the horses. It is his only real vice besides food, and it is a small one at that. Five- and ten-dollar bets mostly, sometimes less.

I help him from time to time, especially when I can inspect the horses. I seem to have a knack for picking winners without the aid of complicated racing forms.

The first time we went to a track—in Lexington—I studied the field and took a fancy to a three-year-old coal-black filly named Wild Lucy Black. She was a frisky animal, full of spirit and spunk. Never mind that she was a 30-to-1 shot.

Ottie laughed when I suggested she was the one to bet, humored me, gave me five dollars, and sent me to the window. I came back a winner, and a $155 dollars richer.

I've always loved black horses. For as far back as I can remember.

Betting on horses is not the surest way to supplement one's income, and some people might wonder why I don't get a job to help support the household. But the reality is: I have one. More than one, actually. I do most of the chores around the house, help with Ottie's paperwork, and take care of my husband. Ottie, meanwhile, likes to say he uses his head to help the family.

"I can't go out and dig a ditch," he says, "but I can use my mind."

Sometimes, he'll even make light of the situation.

"She married an old, crippled man," he once told a person. "Now, she's got to support him."

Deep down, though, Ottie, a once-active outdoorsman who welded on off-shore oil rigs in the Gulf of Mexico and cut trees in the Pacific Northwest, feels awful that he's not able to do more. It is a source of great internal stress. An all-too-frequent gnawing at his pride.

And in my "Field of Dreams," I am compelled to do what I can to ease his pain.

Seventeen

A greeting of love and hope you can obtain eternal salvation. Oh, come home and make peace with God and the church. . . . The man that you have left with is an adulterer and is on the wide and broad way, but you can come back if God moves you.

—Letter from Dad

Rev. James Bettermann is a handsome man, one of those too-handsome-to-be-a-minister kinds of people. And he has a mustache of all things. What the Amish would say about that.

But it was clear from the moment I first saw him, first met him, that he was a Christian of the first order; that he not only believed in God, he worshiped him.

It showed in the red and white sign just inside the entrance to Holy Trinity Lutheran Church:

"We are here as a church to win the lost for Christ and help them and each of us grow as his responsible disciples."

It showed in the inclusive messages in the weekly church bulletin:

"We extend a warm and hearty welcome in the powerful and precious name of Jesus Christ to all who are worshiping with us today. . . . May our Lord bless your worship with us today and may he go with you as you leave to witness and serve."

And it showed in Rev. Bettermann's compassion when he hugged parishioners, and in his easygoing, love-all-people manner of speaking.

He'll tell you he's that way because he wants the Christian experience to be "as real and as concrete as possible. When you show love, when it's expressed, it's more concrete, especially if it's done without ulterior motives.

"I want people to know that God cares about them—and that the pastor cares about them."

It wasn't always this way for him. In the sixties and seventies, he'll tell you, he ingested enough chemicals to last a lifetime. He was part of a drug scene that ultimately would lay waste to many a great mind and many a precious hope. His was a soul desperately searching for meaning—until he found God.

"I woke up one day," he says, "and realized God hadn't quit on me because I was wild. And then I knew I couldn't quit on him.

"The words in the Bible aren't stories to me. They are real, and I believe in them. That's the A to Z motivation for me."

We arrived at Rev. Bettermann's doorstep through Faye, who belonged to the church and taught Sunday school there. Though we had continued to pray in private, having a public anchor for worship was something we desperately needed.

In the back of my mind, I also hoped that the church would one day lift the Amish ban. It was a silly concern when you get right down to it. The Amish would never honor such a move, and it would therefore be little more than a symbolic gesture. But I also hoped that perhaps, with the ban lifted by the Lutherans, it might be easier to visit my mother—if and when such an opportunity arose.

Rev. Bettermann knew little about the Old Order Amish when Ottie and I first began attending his church in Bowling Green. My head covering, he thought, was a bit odd. And I was as shy as they come, slipping in and out of church with barely a whisper.

But over time, as we increasingly sought him out to talk about our ordeal and my upbringing, he came to know the Old Order Amish traditions, and I came to realize there were more than just the two segments professed by the Amish—the Amish and the evil world. There was a third faction. Christians of the world. And they, to my surprise, preached about nonbelievers.

We began going to public Wednesday night Bible study classes conducted by Rev. Bettermann, and we also met privately with him. I was apprehensive at first, expecting all of my Amish learning about the English to bear true. But instead, I discovered a man who preached just as passionately about God, who followed the Bible in all its glory and wisdom, and who was more interested in the word than the denomination.

It helped, of course, that Faye was a member of the church, that Ottie had been baptized Lutheran, and that I had joined the church choir. But more important was that we acted like Christians in all that we did. With the Amish, it was the Amish first. With the Lutherans, it was God first.

Faith to the Lutherans was a relationship with God, not with an institution.

After being somewhat adrift in a sea of abandonment since the ban, I began to feel a sense of belonging again, and Rev. Bettermann came to be a trusted friend whose views on religion coincided with those I had clutched dearly for so long.

"Christianity," he'd say, "is about what's in your heart. I'll concur that the Amish are terribly religious. But I struggle to see how they're Christian, because they are so legalistic, they are so works-oriented. We may use the same terminologies, but they don't mean the same things.

"Unconditional forgiveness, for example. There isn't anything in the Amish system that gives me any picture of unconditional forgiveness. Everything is tied to behavior, and that moves it dangerously close to being outside the Christian realm."

One of the first discussions we had with Rev. Bettermann was the concept of grace—that it is a free gift from God, no strings attached, and that God loves and forgives us unconditionally.

We also talked about how divorce was not an unforgivable sin. About how my marriage to Ottie was okay in the eyes of God and would not banish me to hell.

It took me awhile to truly believe he wasn't just saying such things to make me feel good. That he wasn't simply telling me what I wanted to hear.

But once I was convinced, I couldn't wait to ask him the question I'd held in for months.

"Can you lift the ban?" I asked.

"Yes, I can," he said. "It won't be recognized by the Amish, but, yes, I can."

And so, on June 8, 1997, a year to the day after I left the Amish, Rev. Bettermann made plans to lift the ban.

I awoke early that morning, filled with a sense of excitement and anticipation. As much as I had tried to hide my feelings—in the stoic way of the Amish—I had broken down on several occasions. It was embarrassing, this weakness of mine, but I couldn't seem to stop it. Pain has a way of bubbling forth, no matter one's resolve, and my pain was deeper than I had allowed.

Ottie would hold and comfort me during these times of despair, making it both more tolerable and more intense. Because the more sympathy he showed, the harder I wept.

Sometimes, to spare him the agony of my grief, I would hold in the tears until I ran an errand or until he went to sleep.

It wasn't that I wanted to return to the Amish, or that I was unhappy. I merely missed my family.

Knowing that, Ottie would often offer to call neighbors of my parents in Iowa so I could talk to my mother, although each time I declined. I didn't let on why, but I can say now that I didn't want to bother other people and, strange as it sounds, I worried my father would think it was a misuse of the phone.

Ottie was then—and is now—my rock, but on this special day of absolution I would have to forge ahead without him. He was laid up in bed with a serious case of gout.

I drove the thirty-five miles to Bowling Green, alone in the physical sense but joined in spirit by my faith in my husband, pastor, and Savior. And when I got to the church, seeing

Rev. Bettermann and the congregation I had come to love further enveloped me with conviction.

The defining moment came at the close of the 10:30 A.M. sermon, which was based on 2 Corinthians 4:18: "While we look not at the things which are seen, but at the things which are not seen: for the things which are seen are temporal; but the things which are not seen are eternal."

Rev. Bettermann led me to the altar and told the congregation what he was about to do. The room was packed. Three to four hundred people maybe. The modernistic blue, green, purple, and white-stained glass on either side of the altar glowing bright in the morning sun.

"She has not only been rejected by her family, but told she is eternally damned by Christ," he told the parishioners. "We want to extend to her our love and our friendship and our acceptance. And we want to lift the ban."

"Yes," several people said in unison. "Amen."

"Those who agree that this should be done should show your support by raising your hands," he said.

Nary a person abstained.

Rev. Bettermann would later say that as he carried out the ceremony, he suddenly was moved in a way he hadn't imagined possible.

"It was more emotional than I'd thought it would be," he said. "I didn't think about it ahead of time, but then I got caught up in the moment."

So did I.

I was nervous and comforted at the same time, drawing

strength from God and the loving eyes and voices of the congregation before me.

When Rev. Bettermann uttered the final words—". . . the ban has been lifted and we accept Irene into our fellowship"—some of the parishioners wept, all of them applauded, and many came forward to hug me.

"If there's anything we can ever do for you," many of them said, "let us know."

How wonderful it was to be embraced by so many people, strangers some of them but friends in Christ all of them.

Four days later, Rev. Bettermann sent a letter to my uncle and former bishop, Elmer T. Miller, notifying him of the church's action:

Dear Mr. Miller:

Greetings in the name of our Lord and Savior Jesus Christ.

This letter is sent to inform you that this past Sunday, June 8, 1997, we received into our fellowship Irene Miller Garrett.

It is our understanding that she has been placed under the ban by your church for reasons with which you are most familiar. As a Christian fellowship, our congregation voted to rescind said ban and to accept Irene as a member in good standing.

I am aware that our action may not meet with your approval, but I believe what has been done is both pleasing to the heavenly Father and a blessing to Irene. Out of courtesy and concern, I share this with you on behalf of the people of Holy Trinity.

Sincerely,

Rev. J.A. Bettermann

My uncle never responded to the letter, not that we expected him to. But at the very least, the lifting of the ban had served a purpose for me. It had given me a temporary peace of mind. It had confirmed what I'd known all along about God.

And when I left the church that day, the landscape around me at once seemed more striking and vivid. The blue in the sky was bluer, the songs of the birds more robust, and the leaves on trees sharper, more clearly defined.

Another meaningful thing also happened that day. I wore my full head covering for the last time. Of all things Amish, the head covering was one of the hardest to part with—as difficult, certainly, as shedding Amish guilt. I had worn it every day—every night—of my life. It had become as much a part of me as my own skin and was at the core of my spirituality.

But on that day, I quietly, unceremoniously, retired the head covering to a box in my closet, where it now rests with my other Amish clothes.

Sometimes, people will ask about my Amish attire and I'll pull out the old stuff for an impromptu show-and-tell. I'll even let people try them on if they're interested.

But that's not the only reason I keep them. Ottie says I should have them to show our children so they'll know about my heritage and, in turn, theirs too.

I look forward to the day when I can do that.

Eighteen

(Lasz das blut Christe dich reinichen als der schnee. Wir hoffen der feind sein macht wird genommen und dasz du zurück kommst.)
Let the blood of Christ cleanse you like the snow. We hope the devil's power will be taken so that you come back.

—LETTER FROM DAD

I'm not sure our lives ever really slowed down after the first year. There were breaks along the way. Very short ones. Then things would speed up again.

The rhythm had nothing to do with the letters from Iowa. They remained constant and nagging, and over time began to chip away at the euphoria I felt after the ban was lifted.

For the moment, though, we were too busy with other things to dwell on the negative.

Ottie and I were about to become stars. Public figures, anyway.

A conversation with a former Amish woman in Kentucky gave Ottie and me the idea of compiling a book of true stories about people who had left the Amish. Because the Amish are a people of few words, we thought these testimonies would be enlightening to others in the outside world. And though our intent was not to condemn the Amish, we hoped it might give people thinking of leaving the fold the courage to fulfill their dreams.

We formed a corporation (Neu Leben, Inc., which means "new life" in German), got the names of former Amish through groups that annually hold X-Amish reunions, and traveled to Indiana, Ohio, Pennsylvania, Minnesota, Missouri, and Iowa to conduct audio and video interviews.

Not everyone we contacted wanted to participate; even ten to fifteen years after leaving, some X-Amish were concerned their comments might adversely affect relationships with family members still among the Amish.

But we gathered about twenty-five testimonies, returned to Kentucky, winnowed the list to ten, and hired Western Kentucky University students interested in being published to help do the writing.

The result was *True Stories of the X-Amish,* a 124-page series of essays about real people. It was accompanied by forty-three of Ottie's photographs.

Rev. Bettermann wrote the introduction, which served to set the tone for the book:

> *There is one word which serves as the linchpin, connecting all of these stories of the X-Amish, one word which weaves through each account as a thematic backdrop. The word is freedom. Freedom to choose where to live and how to live, freedom to choose what to believe and what to reject. Freedom to make the basic decisions of life.*
>
> *While most of us take such freedom for granted, the stories that follow describe people who longed for yet lived without the fundamental right to choose the course of their living. . . .*

Ottie worked the network of distributors to get the book in stores and on-line. We also contacted media outlets far and wide, seeking stories on the book—the most inexpensive advertising one can acquire.

The response was overwhelming. Not for the book, much to Ottie's chagrin, but for our own story of struggle and success. Every newspaper in every city we went to for book signings did a story. Local television affiliates also jumped on the bandwagon, and later the TV news magazine *Extra* and a French TV station.

Then *Glamour* magazine called. A senior editor said she was interested in having someone at the magazine do a story on me.

"What about the book?" Ottie asked.

"Oh, I'm not interested in the book," she said. "I'm interested in Irene."

Poor Ottie. It was a pattern that would continue for months.

As always, he would keep a game face on and try to find humor in the situation.

At one book signing, he found himself leaning against a wall beside the bookstore's manager, watching as people crowded around me.

The manager asked him: "What do you think about taking a backseat to all this?"

"Well, I don't really mind," Ottie replied, "as long as it's a Lincoln."

Glamour's story would bring him little consolation. The senior editor decided to take the assignment herself, spent three days at our Horse Cave home, then sent a photographer, photographer's assistant, and makeup artist to do the photo shoot. The ensuing four-page spread didn't say much about the book, but it said a lot about our lives.

It was part of a series called "Wow Women," and ran in August 1999 under this headline:

"Escaping Amish Repression. One Woman's Story. Fearing a future of near slavery, Irene Miller fled her family and a sequestered life that forbade her an education, a career and marriage to the man she loved. Here's how she's dealing with her new modern world."

The reaction to the article was almost instantaneous. *The Nashville Tennessean* did a big story on its "Living" cover under the headline:

"An individual is born. Kentucky woman casts off conformity of Amish life, gingerly enters modern world."

Next came award-winning movie producer Beth Polson inquiring if we'd be interested in giving her rights to do a CBS-TV movie based on our lives. Loosely based, it turned out.

Once again, Ottie had to swallow his pride.

"Did you like the book?" he asked her.

"What book?" Beth said.

"True Stories of the X-Amish."

"I haven't even seen the book," she said.

Ottie sighed.

In the end, we sold the movie rights to Beth, even though several other film companies also came calling. Beth's credentials were simply too sound to ignore. Her Pasadena-based

Polson Co. had produced numerous heralded TV movies, including *The Christmas Box, Go Toward the Light, This Child Is Mine, Not My Kid, A Place to Be Loved,* and *Going Home.*

We expected a close rendition of the story we gave the screenwriters, who were two Mennonites from Goshen, Indiana. But when we read the shooting script for "This Side of Heaven," we were shocked. The locations and most of the names had been changed; I was now Rachel Beachy and Ottie was Jack Dunbar. Even more alarming, the script took great liberties with the facts.

We were concerned my family might think we hadn't told the truth to the screenwriters. We also worried that some of the changes might put the Amish in too kind of a light, perpetuating the warm and fuzzy stereotypes portraying them as unfailingly righteous.

But eventually we grudgingly accepted Beth's explanation that the movie would be an entertainment vehicle, not a documentary. And we recognized that the name-altered script, which focused on the love story in the context of "independence versus convention," offered a lot of advantages, most notably protecting, at least for the moment, my parents' identities.

It was also Beth's belief that the movie would not tread lightly on the Amish, but instead provide "an interesting way to look at a simpler life and realize that that's not perfect, either."

That would put *This Side of Heaven* a notch ahead of another Amish film, *Harvest of Fire,* a 1996 TV movie about barn burnings that starred Patty Duke Astin and was filmed in Kalona. Some of the portrayals of Amish life in that movie were simply laughable.

Whatever the outcome of *This Side of Heaven,* the Amish will probably dismiss it as nothing more than English fancy, much the way they dismissed *Harvest of Fire.* The Amish have a decidedly jaded view of film and TV, believing neither can be believed, that shows are made up to sway people in one direction or another, and that there is an evil attached to all of them.

The antenna on the roof is the devil's tail, they say, and the TV in the living room is his tongue.

This ominous tone was conveyed in a letter my father wrote:

> *Irene, I so wish that movie would be stopped, which you probably could, right? Wilbur's neighbor said, "There's a black person in every movie and this time it will be your Dad." I'm not acquainted how these go, but if this is true, I just want to keep apologizing and ask to be forgiven.*

I'm reasonably certain the word "black" means a "bad" person and does not refer to race, although many Amish do hold deep prejudices against blacks and discourage any kind of interracial contact.

"Ever seen a sparrow with a robin?" they'll ask. "Ever seen a bluebird with a cardinal? Well, if animals are smart enough to be with their own kind, humans should be, too."

It's not exactly a ringing endorsement for equality.

Nineteen

Still praying that some way, somehow, the family may be united again, for this is hard to go on.

—Letter from Mom

The best part about all the media attention was not our fifteen minutes of fame. It was the people who, through familiarity with my story, came to terms—or at least tried to—with crises in their lives.

There was the attractive middle-aged woman who came to a book signing and told how she had been harangued by family members her entire life because she had been born out of wedlock. She had been constantly reminded that she was a mistake.

I listened, which is all people really want anyway. Then I tried to help her in my own small way. In the only way I knew how.

"Do you believe in God?" I asked.

"Yes," she said.

"Well," I said, "God doesn't make mistakes, and he made you, didn't he?"

"Yes."

"Then you can't be a mistake."

Tears of joy filled her eyes and the years of torment washed down her face.

"Thank you," she said.

"You're welcome," I said. And I smiled.

There were letters, too. From all over the country. Not only from X-Amish but from English people who'd struggled with their faith to make a go of it in this life. Some of them had even read Ottie's book.

One woman wrote:

Reading your husband's book reminded me of what I went through when I left the Jehovah's Witnesses. I was one for seven years and when I tried to leave, it was very stressful. They also keep to themselves and tell members not to mix with the world, not even family members who are not JW.

When I left, my ex-husband found out where I was living, even though I was hiding. Then he and an elder came and talked to me, trying to get me to see the error of my ways. But I wouldn't go back.

I had it easier than the X-Amish because I didn't have a van load crying and pleading. But it was bad enough. After that, I had a twitch in my eye for several months because I thought God was going to destroy me because Armageddon was going to be here surely any minute now.

Religion and spirituality are not necessarily the same thing. And now I'm happy and free. I bet you are too!

A husband and wife wrote:

Some time ago your story was featured in a Nashville paper, and my husband and I read it with interest. Maybe it caught our interest more so because of our past experiences, as we also grew up in a very conservative church where a lot of emphasis was placed on being 'separated' from the world. . . .

The day came, however, when we felt that we were in bondage and we longed for liberty. Our bondage was the prison house of sin, self, and self-righteousness, wherein Satan had taken us captive. God was so good. In his word he showed us that there is deliverance, freedom, victory if we come to him with all our heart, confessing and forsaking our sin, giving our life completely in his control, believing that Jesus died for such repentant sinners, and through his blood he will set us free. . . .

I love you and want you to have a heart-happiness that is real and lasting. . . .

A twenty-three-year-old woman wrote:

I left the Old Order Mennonite faith a little over three years ago. The question you said you asked yourself—"Am I going to hell?"—is still a question I ask myself every day. It's so hard to get past those feelings of guilt and condemnation that were instilled in us from an early age. I am very angry and upset that my parents put me through that kind of mental hell. . . .

I am so glad to see you were able to go on, be strong and be an encouragement to others that need deliverance from a life of bondage by man and not a life of living for God. . . .

The woman went on to say she had recently wedded an English man, and that she had put off marrying him because she was worried about the complications it would cause with her family. I couldn't help but think about my own situation.

One of the more heart-wrenching letters came from a forty-one-year-old Tennessee woman who'd recently left the Amish. She had never been married and possessed only an eighth-grade education, but she was valiantly trying to better herself.

She wrote:

> *I'm not a good educated writer but I hope you can understand my writing. I enjoy do [sic] the G.E.D. schooling. I learn a lot. I have come a long way since I start [sic] going to the adult education classes. Lots of the English people around here attend the classes. The classes are given in A.M. and P.M. twice a week. I mostly study at home. I go to class when I need teachers [sic] help.*

That one made me just a little bit angry.

Here we were at the dawn of the twenty-first century and we still had people in dire need of education, in desperate searches for healthy balances in their lives.

For these people, the clock had, at least for a time, continued to march on without them. National unemployment was on the verge of reaching a thirty-year low, John Glenn had just returned to space aboard the shuttle *Discovery*, and Doctors Without Borders had been awarded the Nobel Peace Prize for helping the world's sick, injured, and impoverished.

Yet right in our own backyards, we had people being smothered by tradition, doctrine and bias.

What was wrong with us? I wondered. How was it that we had arrived at this point in the road?

That we could help some of these people was great solace, and some, upon hearing our story, arrived at our door seeking assistance. We didn't go looking for them, mind you. We merely made ourselves available.

And we went right to work when an Old Order Amish couple and their eight children came calling. They lived in squalor in a tin-roofed, two-room house on an English farm in Kentucky. The father made three dollars an hour (never more than eight thousand dollars a year) working the land, which isn't anywhere near enough to support a family that large. Worse, he had gotten himself in some kind of trouble with the Amish. They had punished him for using electricity in the home; had unplugged a freezer, spoiling all the family's food, and had wrongly accused him of drinking alcohol.

When we first encountered the family, we were horrified at their living conditions. They slept on mattresses on the floor, or sometimes on the very floor itself. Their clothes were so soiled that even several washings could not remove the stench, and the children were covered with lice.

To help them leave the Amish, we knew we'd have to find accommodations in another state. So, Ottie called one of the subjects of *True Stories of the X-Amish* who now ran a metal fabrication plant in Alabama, and the man graciously bought the family a mobile home and hired the father on at the factory for eleven dollars an hour.

That was several years ago, and the family is adjusting well. One of the girls even became homecoming queen.

Twenty

If you should die now, where would you spend eternity? One of Satan's famous tools of this day is to mix world with religion. He wants you to believe that as long as you pray to God, you may live and do as you wish.

—Letter from Wilbur

When we weren't tied up with screenwriters, book signings, and the like, Ottie and I would spend time at home watching television, gardening, baking, entertaining friends, visiting his family, going out to eat, or sightseeing.

Our little white Maltese, Fluffy, also kept us on our toes, demanding attention—and giving it back tenfold—incessantly. Sometimes, we'd also work on getting our parakeet, Peaches, to retreat from his silence by playing bird songs for him on the CD player.

I would be frequently reminded of our journey by my crystal swans in the living-room curio cabinet, and by the Kentucky Old Order Amish, who would steer their buggies by our house on Old Dixie Road to glimpse the now-famous renegade couple.

Occasionally, I would get out my blue velvet Amish hymn book to sing a song or two, including my favorite, "I Need No Mansion Here Below."

I would also remember lighter moments, like how Ottie used to tease the Iowa Amish with word games—when he wasn't frustrating them with serious debate.

One game went like this:

Say shop three times.

Shop. Shop. Shop.

Now say it five times.

Shop. Shop. Shop. Shop. Shop.

Now say it three times quickly.

Shop, shop, shop.

Now say it once slowly.

S—h—o—p.

What do you do when you come to a green light?

Stop.

So fixated were they with the word "shop" that they were determined not to say it and would mistakenly reply "stop." It worked every time.

In my spare moments at home, I also began reading scriptures in the Bible I had never paid much attention to growing up. Scriptures the Amish choose to ignore. Scriptures that don't coincide with their traditional thinking about such things as going to war, focusing on works, and salvation.

1 John 3:16 addresses the war issue with this passage: "Hereby perceive we the love of God, because he laid down his life for us; and we ought to lay down our lives for the brethren."

There is a similar message in Jeremiah 48:10: "Cursed be he that doeth the work of the Lord deceitfully, and cursed be he that keepeth back his sword from blood."

On the matter of grace and works, Ephesians 2:8–9 clearly spells out the importance of grace: "For by grace are ye saved through faith; and that not of yourselves. It is a gift from God: Not of works, lest any man should boast."

And regarding the Amish belief that salvation isn't assured believers, 1 John 5:11–12 leaves no doubt: "And this is the record, that God hath given to us eternal life, and this life is in his Son. He that hath the Son hath life; and he that hath not the Son of God hath not life."

My entrance into the modern world had not only accorded me the benefits of convenience, it had given me the freedom to read the Bible as it should be read. To understand the full breadth of its teachings.

I have not absorbed all of it yet, not completely. And there are still some biblical issues that conflict with my desire to keep an open mind.

The whole concept of dinosaurs living sixty-five million years ago, for instance, seems incongruous.

The Bible, which only goes back six thousand years, mentions dragons but not dinosaurs. So how is it that archaeologists find the bones of dinosaurs but not the bones of dragons?

Moreover, the Amish are convinced history only goes back eight thousand years. How, then, can people be sure the bones they're finding are millions of years old? And how can they be sure they're putting the right appearances on the skeletons?

This all came to a head one day in our living room when Ottie and I met with Rick Farrant, my coauthor, his wife, Susie, and their young daughter, Amber. I think Rick started the debate by mentioning Chicago's exhibit of the world's largest Tyrannosaur, and the rest of us joined in.

"They say some of these dinosaurs are sixty-five million years old," I said. "Who measured this? By whose measurements were these made? What proof have they got?"

"Well, I know you didn't learn this in school, but scientists have this thing called carbon dating," Rick said. "And through carbon dating, they are able to determine how old bones are."

"But how do they know they're putting the right appearance on this so-called dinosaur?" I said. "The only thing they've got is the bones. What if you find a dog carcass and put a cat covering on it? They could do that, couldn't they?"

"They could," Rick said. "But scientists, while not 100 percent sure, have many techniques for determining what a dinosaur might have looked like."

Rick then raised this question: "How do you know that the words you read in the Bible are the exact words passed down through the centuries?"

"Oh, don't go there!" Ottie bellowed.

"I'm not trying to shake your faith," Rick said, looking at me.

"And you won't!" I fired back.

We all had a good laugh, and I acknowledged that little nuances had likely been altered in the numerous translations of the Bible, but that the general messages had been left intact.

Ottie, in my support I suppose, then mentioned that some Amish don't believe man has been on the moon—that man could have faked it on television, just as they do with movies.

Maybe, he offered, it's not so unusual for some people to think man faked the dinosaurs, too.

For some reason, that reminded me of an Amish man who thought all the waters from the Great Flood were deposited in space, hence the blue skies. If man had truly gone into space, this person reasoned, the rocket would have penetrated the water and caused another flood.

I don't believe that for a minute, and I'm positive man has been in space. But this dinosaur thing—well, that's a whole different matter. Among other things, it seems to challenge God's handiwork and support the theory of evolution. God created man, plants, animals, sky, water, everything in seven short days. He didn't need sixty-five million years. In fact, one could presume from reading the Bible that there's no such thing as a span of time that long.

"But what if," Ottie said, "God's sense of time is quite different from ours? A day is man's creation. It may have been one day to man and quite another to God.

"So, at the time of creation, maybe there was no time. Because God is infinite. So to God, there is no time."

"But I don't believe," I said, "that God had to depend on evolution to take thousands and thousands and thousands of years to make a little bird."

"Did God tell you he made everything?" Ottie asked.

"Yeah."

"Did God tell you how he made it?"

"No."

"Couldn't he have used evolution, then, to do it?"

"Yeah, he could have. But he didn't have to. I don't know whether he did or not."

"But couldn't he have used evolution?" Ottie persisted. "Because the Bible doesn't say how he did it. It just says that he did it. So he could have said, 'I'm going to create all these animals, but I'm going to do it in such a way that everything works itself into the perfect state that I want. So we have creatures crawling from the oceans to the land, then walking, then dinosaurs and lizards and flying reptiles, and little horses that become regular horses, and then there's man."

This was a concept I hadn't thought of before. A very strange, uncomfortable notion.

But as difficult as it was to fathom, I knew I must consider it—for the sake of keeping that great virtue called an open mind.

"Well," I said, "I'm not going to be like the Amish and say there weren't dinosaurs, because there might have been."

Emphasis on the *might,* that is.

Later in the day, the subject was revisited when we drove south to Meshack Creek in Monroe County. We stopped where a narrow country road ended at water's edge and got out to inspect long, deep grooves in the creek's hard-rock bottom. The marks were several feet apart and ran diagonally against the current from one side of the creek to the other. Wagon wheel tracks, Ottie speculated. The vestiges of a route heavily traveled more than a century ago.

As we pondered his theory and imagined lines of settlers crossing the languid creek thicket to thicket, Rick discovered a veritable gold mine of fossils in the rocks along the creek's banks. In short order, we were all bending over trying to outdo each other in finding the best specimens.

"You realize," Rick said, "that some of these fossils could be as old as 400 to 500 million years. Before the dinosaurs."

I looked at him and nodded, not wanting to get into another protracted conversation. But more to the point, I was so fascinated by these curious little forms that I wanted to collect as many as I could, as fast as I could. There must have been fossils in Iowa, but I can't say that I ever noticed them.

I later learned that scientists believe Kentucky was once a shallow, prehistoric tropical sea, and that what I held in my hands were the remnants of tiny sea lilies called crinoids (or Indian beads), squidlike animals called cephalopods, shell creatures known as brachiopods, and corals and sponges.

I returned home with a plastic bag filled with spirals, cylinders, cones, and shells, and deposited it on the front porch for a more detailed examination later.

"Don't lose those," I told Ottie. "I'm going to find me enough of these fossils that I'm going to build me my own dinosaur."

I said it half in jest. Emphasis on the *half.*

Twenty-One

Are you ready to meet Jesus when He comes? . . . Are you making the scars in Jesus' hands bigger if you keep right on going your way?

—Letter from Bertha (Sister)

Bertha, my older sister by six years, had promised me when I left Kalona that she would write frequently to let me know how things were going with my family. Once every two weeks at least. Maybe even once a week.

But since my departure almost four years ago, she had written only three times.

I would later learn that she had lost her desire to write because my father insisted on reading every letter sent to me. It didn't help that she is as timid as they come.

Whereas I was determined not to be pushed down too far when I was growing up, Bertha lost the battle early on. She was constantly made fun of for being overweight, for having trouble with her feet, and for having bad eyes. By my family. By others.

Not that I was small or didn't have my own physical ailments.

I weighed 170 pounds when I left the Amish. But for an active, big-boned woman like myself, it was a healthy 170. I could do almost anything a man could do on the farm, and certainly much more than Bertha was capable of doing.

Like Bertha, I also had trouble with my eyes. But I faked it. While Bertha was wearing glasses, I was telling people I could see things I couldn't see.

A remark my father made to me once had me convinced wearing glasses was disgraceful. One time when I told him I couldn't see something, he said in condescending fashion, "Awww, you just want to wear glasses, don't you?"

From that moment on—until I began teaching school—I was determined to make do seeing things with fuzzy edges, or not at all.

I am now trying to undo the effects of indulgent experimentation with English food, a lack of regular exercise, and a happy marriage. Since leaving the Amish, I have put on a considerable amount of weight, so much so that I'm hesitant to reveal how much.

I wish I had stopped the slide earlier, but it's my responsibility. I walk two miles a day and survive on what amounts to a diet of bird feed. My determination and self-esteem will restore me to a farm-girl-healthy 170 pounds. I'm sure of it.

And that is one of the biggest differences between myself and my sister. I do not like to fail, and when I was among the Amish I worked hard to make sure people respected me. My naturally outgoing nature also helped keep the wolves at bay, and my private study—absorbing any outside reading material I could get my hands on—gave me an intellectual edge that kept others off guard.

I wish I could say it was all planned, but in truth these were inherent gifts. I had my destiny. She had hers.

Not that she enjoyed her place of darkness. She once told

me how people belittled her—from the owner of the Stringtown Grocery who paid her just a dollar an hour for a year and a half before bumping it a whopping fifty cents to the teachers who refuse to give her rides back and forth between work and home.

Bertha teaches at an Amish school in Milton, about a two-hour car drive south of Kalona. Her fifteen dollars-a-day salary prohibits her from returning to Kalona regularly, so she boards with a local Amish family. She could become part of a car pool with teachers in nearby Bloomfield, about fifteen miles away, but the teachers tell her it's too far to travel to pick her up en route to Kalona.

She complains about these things, but there is also a helpless acceptance that further pushes her into a submission that is evident in every one of her mannerisms. She walks huddled and stooped, as someone might do fending off a subzero blizzard. She speaks softly, as people often do in libraries. And when she shakes hands, she does so limply and without conviction.

She is, in the words of the Amish, a sorrowful old maid who is destined to remain unmarried—unless a widower chooses to have her as a caretaker for him and his children.

So entrenched is she in yielding to oppression that she is among my harshest critics at times. And clearly, a frightened one. Such was the case in one of her letters:

My Dear and Only Sister Irene:

Lonely greetings sent your way.

How do these lines find you? Since the last time you were at home, we often have to think about you! My thoughts have been I wish I could talk with you again, because I wouldn't mind asking you some questions, but it's not safe to write them down anymore.

Mom and Dad don't know I wrote, but I thought I should write you and tell you I got your letter. . . .

Do you have all your questions answered? Are you ready to meet Jesus when he comes?. . . Are you making the scars in Jesus' hands bigger if you keep right on going your way?

Your only, lonely sister:

Bertha

I would dearly love to help her leave the Amish—to give her a chance at a fruitful and rewarding life. Because inside, she is a wonderful, caring person with a heart of gold who wrote this to me while I was still living in Kalona: "You're the best sister a friend or sister could have."

Now that I'm gone, though, I worry that if I am too direct or forceful in my efforts to sway her, she will stiffen and back away. Perhaps forever.

So I try to tell her how happy I am, and how one can live this new life without abandoning God and without risking the chance of going to heaven. In that way, I hope, she will summon the courage to take a bold step.

Dear Sister Bertha:

In God have I put my trust: I will not be afraid what man can do unto me. Psalm 56:11.

The sun is shining today and it's a beautiful day. It rained on Monday, so hopefully towards the end of the week I'll be able to plant my garden.

I don't feel really perky as I got a sore throat and a cold, so I decided to write some letters. I'm sending your letter to the Milton address because you would still be there. I used the Iowa Amish Directory to find your address, so hopefully I have the right one. . . .

Are Mom's legs doing better? My suggestions may not go far, but for Mom's sake I wish she would go see a specialist if they don't get better. I worry about her health and I'm afraid she could get blood poisoning and lose her legs if she's not careful.

My life is so free, happy, and peaceful, you couldn't imagine the difference, Bertha. My church is a loving Christian church and I've made many good friends. They love me and are friendly to me all the time. Nobody can imagine it, unless they live it or see it for themselves. I wish I could just come get my whole family so they could experience it. What a happy family we would be!

I realize there may be little hope for that, but it doesn't keep me from praying for it, because with God all things are possible. I didn't think I could ever leave, but he was leading me and without his help I couldn't have done it. The only thing that clouds my happiness is any pain it brought to anyone. It would really hurt me if I knew anyone made it harder for you, Mom, or anyone in my family because I left. You're my only sister and I love you dearly. I'll always be there for you or anyone else that needs me. . . .

Love always,
Irene

I find it hard to believe that Bertha will ever leave the Amish, both because of her shyness and because of the Amish propaganda.

The Amish tell their flock that English churches preach about heaven but not hell. They also say English churches twist scriptures to suit whatever needs come down the pike.

Finally, they tell members to beware of false prophets—and that would include people like myself who have left the Amish for the English world.

It is incomprehensible that they can form such opinions without experiencing something first. It is equally disturbing that they feel so compelled to rule by intimidation.

At one time, it would have been inconceivable for me to consider the Amish a cult.

But now I understand the meaning of the word, and I think it's possible they may be just that.

I don't hate the Amish for their transgressions. I feel sorry for them and worry about their spiritual welfare.

I feel sorry for Bertha, too, and pray that one day she, along with my mother, will be delivered from their living hell.

Twenty-Two

(Du weiszt das keine hoffnung ist für ein gebanntglied. . . . Unser wunsch ist das du den Schöpfer suchen tust ob es zu schpot ist. Und der einzigste weg das wir sehen könnten das das getan sein kann ist zurück kommen wo du abgefallen bist und busze tun das übrige deines leben.)
You know there is no hope for a banned member. . . . Our wish is that you seek your Creator before it's too late. And the only way we see this can be done is come back to what you have fallen from and repent and be sorry for the rest of your life.

—Letter from Earl Miller (Uncle)

My learning continued at an accelerated pace, both through travels Ottie and I took, and through achieving a dream I could only obtain in the English world—getting a General Equivalency Diploma.

Both came easily.

Ottie loves to drive, of course. It is the one way he can achieve comfortable mobility. And I love riding with him, taking in the sights, meeting people, absorbing all that is new.

I am no longer as fascinated with the van as I once was. It is now like an old suit. Familiar and reliable, with pictures of me and Ottie propped on the dash, I in Amish attire and he in jeans and a flannel shirt. In the middle of our pictures is a passage from Psalm 119:73:

"Thy hands have made me and fashioned me: give me understanding, that I may learn thy commandments."

We love taking trips to the country around Horse Cave, all the while listening to songs on the radio and holding hands. There are few moments more divine in our lives than sitting in the van, holding hands and watching the sun rise or set—God's spectacular signatures on a day.

We have our other favorite places we like to go, too. Key West is one of them. The Smoky Mountains another. And then there's this little sliver of paradise in the middle of nowhere—Kentucky's Marrowbone Valley. Locals call it the "Marribone," no "w's." (There's something about "w's" in Southern speak; people don't pronounce them.)

The valley on Highway 90, southeast of diminutive spots in the road like Eighty Eight and Summer Shade, is rich with arrowheads and sparse with people, most of whom have decided to leave the hustle and bustle of big cities and forge a quiet life of simple pleasures. There's the young couple with ten children who live off the land. The one-time police officer. The retired Air Force enlistee.

It is here that we would like to live, too. On a plot of overgrown pasture next to the river. In a cabin or a house or both.

We go to the valley frequently—to dream and plan. There's one particular 4.2-acre site we have our eyes on that fronts the highway, has a little dirt access road, and claims part of the Marrowbone River. We've talked about clearing the land first—so we can see it, and so we can retrieve any fossils or Indian artifacts.

Then we will build. And garden.

The southern Kentucky region has long been a favorite of Ottie's. He was born there and, even though he moved to Toledo, Ohio, a year later, he returned to visit relatives in the summers of his youth.

When he was ten, Ottie and three cousins began exploring unmarked caves, including a large one behind the Glasgow City Jail. Inside the cave, in a spacious cavern, they found schools of blind fish no longer than a person's pinky, and planks and chains that once confined slaves. With a bit of luck, they also found their way out after one of the boys dropped the group's only flashlight and rendered it momentarily useless. When they finally retrieved it, searching on all fours, they knocked it a time or two and the light kicked back on.

The boys also delighted in going to the country store at the Etoile crossroads, buying bottles of Coca-Cola and bags of peanuts and, after swallowing a quarter of their beverages, plunking the peanuts into the bottles. It was, for them, just the right concoction of drink and food to pass the time under a shade tree. That, and Moon Pies.

Sometimes in the swelter of those 1950s summers, they would also tie sewing thread to the legs of June bugs and watch the big green insects cut circles in the air—until the bugs' legs snapped off, freeing them from their bondage.

When he wasn't with his cousins, Ottie would go squirrel hunting behind Grandmother Garrett's house, lugging along her .22-caliber single-shot rifle. He wasn't much of a shot back then, and he was much too fascinated with the squirrels cavorting in the trees to ever lift the gun to his shoulder.

Later, he moved with his mother Ersie, brother Benny, and sister Faye to Attica, Indiana, after his parents divorced, grew into a strapping high school defensive tackle for the Red Ramblers, and charmed the girls at sock hops with his dancing technique, learned at the hands of his mother and sister, who taught him for an hour every Saturday when Dick Clark's *American Bandstand* aired.

His mother, who died of cancer in 1992 at sixty, was a fan of Elvis and the Everly Brothers. There was also room in the home for the likes of Patsy Cline and Johnny Cash.

These stories of English life helped form the basis for my education, although I realized purposeful learning meant so much more than absorbing folklore. I needed to get my GED.

I took a sample test at Barren County High School in Glasgow in the summer of 1999 and was told I'd done well enough to skip adult education classes. Only my math scores were dubious, and I took home several floppy discs containing sample algebra questions so I could study them.

On October 6, 1999, with little preparation, I took the real test—and passed.

I finished in the 89th percentile nationally in writing, 74th in social studies, 71st in literature and arts, 64th in science, and 44th in mathematics. It was, I must admit, a pretty good showing for someone who'd only had a formal education through the eighth grade—and a somewhat limited education at that.

All I can figure is that my hours of private study growing up and my year as a teacher gave me enough of a push to be successful.

In any event, I was proud.

I received a letter from the Kentucky Department for Adult Education and Literacy confirming my feelings, and my accomplishment:

Dear GED graduate:

I would like to congratulate you on passing the GED test. Your high school equivalency diploma is enclosed. I am honored to have this opportunity to reward your outstanding effort.

Earning your GED marks an important milestone in your life, and your success is an inspiration to students across the Commonwealth who are currently enrolled in a GED program. . . .

Congratulations once again on attaining your GED! I encourage you to continue this positive momentum as you pursue your future goals, and I wish you the best in all your endeavors.

Sincerely,

Reecie D. Stagnolia

Acting Commissioner

The formal graduation came on May 15, 2000, in the Barren County High School Auditorium—on the heels of President Clinton traveling the country, including Kentucky, in support of school reform.

I wore a royal blue cap and gown and a smile the size of the Bluegrass State that night as I waited my turn among the ninety-two graduates. Someone sang a couple of gospel songs and a parade of speakers made sundry announcements, including one who mentioned something to the effect that 73 percent of high school seniors wouldn't be able to pass the GED. A woman sitting next to me leaned over and whispered, "Man, she just made me feel smart."

I was feeling pretty bright at the moment, too.

One of the speakers had told me earlier she was going to mention that I was once Amish. But when her time came, she became flustered and forgot. She later apologized to me in a letter, but I held no ill will. The important thing was that I had graduated.

After the ceremonies, the celebration continued at a party at Faye's house attended by friends and relatives. There were soft drinks, cheerful banter, and a cake that pronounced: "Congratulations Irene 2000."

It got me to thinking about going to college one day to ensure my future, perhaps to be a nurse or a youth minister. I like the idea of helping people, although I'm a little hesitant about being around injured people day after day.

I once told Ottie: "If somebody comes into the hospital one day with their leg wide open, I'm not sure I want to be exposed to that."

Somebody told me I'd also have to witness an autopsy as part of my training, and I'm not sure I want to go through that, either.

There's a third career option that has some appeal: professional photographer, specializing in weddings, reunions, and portraits of families. With Ottie's guidance, I have grown to enjoy taking pictures and even shot one of a restaurant fire that was published on the front page of the *Glasgow Daily Times.*

Perhaps because I have no photographic images of my own childhood—something I deeply regret—I like the idea of creating memories for other people. A person can remember so much more through pictures, and they can derive years of pleasure simply by taking them out and reminiscing every so often.

As I do with the pictures Ottie has taken of me.

Perhaps one day I'll be able to put pictures of my children in our photo album. Making babies has become a priority of ours—even more important than my career.

The first two years out of Kalona, we had to be careful, because I hadn't had any of the vaccinations English children normally get during their adolescence. The Amish are reluctant to get children vaccinated—because they distrust the English, because unfounded rumors have spread about deathly allergic reactions to the shots, and because the Amish are wedded to various home remedies.

The Amish have a variety of concoctions they use to "cure" everything from acne to heart trouble—and in the case of my mother, open sores on one's legs. They're passed from community to community through Amish newspaper articles in *The Budget* or *Die Botschaft*.

I can't vouch for many of the remedies. They're simply too odd. Like ingesting nine plump, steamed raisins a day to clear up acne. Or using strange-sounding products like Watkins Petro Carbon Salve (for corns), Silent Nitezzz (for snoring), and a light-green, slippery Chickweed Healing Salve

("Good for all skin disorders; skin cancer, cuts, burns, and poison ivy"). Or eating cayenne peppers and kelp to calm an irregular heartbeat.

But some of them work; I know from firsthand experience. A honey-and-flour paste does wonders for drawing out splinters and bee stingers.

There's nothing homespun, though, for measles, chicken pox, whooping cough, and the like, and every year in Amish schools across the country there are uncontrolled outbreaks of disease. Fortunately, I contracted only one of them—chicken pox. I'm not prepared to take the same chances with my own children.

I got my first shots at twenty-two—from a pediatrician of all things—and he told us to hold off on unprotected sex for two years to make sure the immunizations had taken properly. And so we did.

But lately we've been wondering—with no pregnancy in sight—whether we'll be able to have children. We've visited doctors and are trying a few twenty-first-century remedies. I am so preoccupied with the idea of having children—two at least, maybe four—that I'm willing to do whatever it takes.

Ottie, who already has four children of his own either grown or living with their mothers, would also like more children, though he worries he might not live long enough to see them to adulthood.

But we're working on that, too. Ottie has begun a diet and exercise program in earnest, with plans to get down to 350 pounds by Christmas 2001. So far, so good. He's lost fifty-three pounds in the first eight months.

If we can't have children together, I suppose we could have foster children or adopt, although both options have their unpleasantries.

It would be hard to get attached to a foster child, only to have it yanked away at a moment's notice. And if I went to an adoption home, I wouldn't be able to pick just one. I'd want them all.

I wouldn't be able to stand the hurt look on the faces of the children who weren't chosen.

Twenty-Three

How could you do such a thing and bring your family so much sorrow? Don't you get lonesome for your nieces and nephews? These have been some of the questions in people's minds and many people are concerned about your soul.

—MARY ANN MILLER (COUSIN)

I had expected that my leaving Kalona would lead to a ban and an estrangement with my family. What I hadn't expected was how the news would spread among the Amish beyond Iowa, eventually seeping insidiously into my life as an English woman in Kentucky.

For a time, it merely amounted to stares of disfavor whenever I encountered Amish in stores or restaurants who had heard of my departure. Some would even turn away.

But what happened at an Old Order Amish grocery near Three Springs, northeast of Glasgow, was a shock of much greater proportion.

I had shopped the store before without incident. On this particular day, however, searching for seed potatoes and kohlrabi, I was confronted by the store owner shortly after I entered.

"Are you Ottie Garrett's wife?" he asked me.

"Yes," I said.

"Well, I can't sell to you then. You know how it is. We just can't do it."

I returned to the van, told Ottie what had happened, and he drove to the front of the store and honked. The owner, fidgeting a bit, came out and approached the driver's side window.

"Hello, Ottie," he said.

"What's the problem?" Ottie inquired.

"You know the Amish way. I can't sell to her because she's in the ban. It's got nothing personal to do with you, Ottie. It is with her that we have an issue."

"But you've served her here before."

"Well, I didn't realize until recently who she was. You know I can't do it, Ottie. You know that."

We paid a visit to the settlement's bishop five minutes up the road, but he was of little help. He talked about how the rule was something passed down through generations—a frequent excuse for Amish inflexibility—and that it was not within his power to make an exception.

"I wish it wouldn't be that way," he said, "but it is."

I could have forced the matter, as some X-Amish have done. I could have gathered my produce and walked to the counter. Because a ban requires other Amish to refrain from taking things from the hands of a shunned person, they would have neither accepted my money nor recovered the produce. They would have let me walk out of the store without paying.

But, of course, I couldn't do that. It wouldn't have been right. It would have been tantamount to stealing.

We also could have sued the owner. In the English world,

it could be viewed as a form of discrimination. But we didn't do that, either.

We have, for the time being, filed it away as yet another example of the rigid, punitive society in which the Amish dwell. Where an Amish storekeeper in Kentucky worries that any lapse in vigilance will bring the wrath of Amish in Iowa.

Since leaving, I've also had chance encounters with people I had known as an Amish woman. Former friends and acquaintances across the country.

I try to avoid such occurrences during our sightseeing travels, which sometimes take us to—or through—Amish settlements. The meetings are too awkward, too strained. The Amish don't know what to say to me, and I don't know what to say to them.

And when the Amish do speak, it's often something like: "Oh, your poor family. What they must be going through."

The prospects of such tension prevent me from moving about freely in Amish areas. A similar tension is also what kept me from my family in the first year. I was afraid to face my father's anger, afraid to witness a further deterioration of my mother's health, and afraid to feel the standoffish behavior I knew some of my siblings would exhibit. I wasn't even sure any of them would consent, without condition, to see me.

For all the times someone has told me I'm courageous, there have been many others when I've been decidedly less so.

Shortly after the ban was lifted by the Lutherans, though, I did make a side trip to Kalona while Ottie, I, and a friend of mine were en route to Minnesota. No one was at the farm that day, and I was later told that my mother, father, sister, and three of my brothers had gone fishing.

Elson and Wilbur, though, were at their respective homes and I visited both. Elson was, as always, cordial and understanding. At Wilbur's, I was never invited in, and I spent the entire time chatting with relatives under a tree. The conversation was stilted and painful.

In one place, I felt like a sister. In the other, I did not.

I once wrote a poem about traveling, not so much for myself but for Amish drivers and their human cargo:

As we go on this trip, O Lord,
And we leave our friends and family dear,
Keep us all in your loving care,
Whether we travel far or near.

Protect us all from misfortunes and harm,
Let thy blessing rest upon us,
Lead us on the path we should go,
With your love, mercy, and kindness.

Lord, help us see your wondrous works,
Which were wrought by your powerful hand,
Rivers, mountains, valleys, and oceans,
Reveal the beauty throughout all your land.

Keep us till we are safely home,
To meet our beloved with joy and peace,
Lead us safely to our heavenly home,
Where love, joy, and happiness never cease.

It was, I thought at the time, a simple, comforting yarn for any hurdles encountered along one's journey. But no piece of prose, no passage from Scripture could prevent the dread I felt before going home, or the bittersweet aftertaste of the visit.

There is a saying: "You can never go home." And in my case, it's true. Although I may be able to physically return, home will never be the same.

I am now an outsider—one of those people I once avoided, and once criticized unknowingly.

Twenty-Four

I miss those egg hunts we had last year. Remember? Those were fun. Even what was on the inside. I also miss those thrilling, exciting, and very fun Pictionary games. That was usually one highlight of school.

—A FORMER STUDENT

Occasionally, a letter would arrive that would lift my spirits. If it wasn't from Elson, it would be from relatives on my mother's side, or from former students of mine.

Like Elson's letters, the students' correspondence were breezy, filled with information about the weather, and school, and pleasant recollections. No preaching. No judgments.

One student wrote to wish me a happy twenty-third birthday, although she didn't quite say it that way. She wrote: "Remembering your birthday."

In a letter spotted with smiley faces, she also told about a harsh winter, a couple having a baby, making paper balls in school, and a student who had to repeat third grade.

Another student wrote extensively about school activities, including a "Tip Toe Day" in which students walked around on their toes all day. Not as a punishment, but as a rather silly game to break winter's doldrums.

She related some of the books they were reading in school, among them *Night Preacher, The Mysterious Passover*

Visitors, and *Along Lark Valley Trail*. And she told how a big hollow tree near the school's basement entrance had been cut down.

Another former student's letter was even more pedestrian. He wrote about someone breaking their arm, people eating pizza and ice cream for a person's fourteenth birthday, and someone trying on a jacket.

The students' letters were fun to read and momentarily made me forget the intense campaign by my family to shame me into returning. If there was one thing the Amish couldn't take away from me—or my former students—it was the good times we had shared in school.

But even those fond memories were not enough to overcome the overly dramatic letters from my parents, siblings, uncles, cousins, deacon, and former friends. It took me days, sometimes longer, to rouse myself from the funk that would set in after their letters arrived. It became apparent that the lifting of the ban had not been the cure-all I had hoped for.

One cousin shared this story she had found in an Amish magazine after it was brought to her attention by a minister:

> *A certain man went hunting. He had a helper with him. They came upon a flock of wild ducks. The hunter shot into the flock, and a number of birds fluttered to the ground.*
>
> *"Quick, go get the crippled and lame ones," the hunter said to his helper. "Don't worry about the dead ones for now. We know we have those."*
>
> *The minister went on to say that is how Satan works. He is out to get those who are spiritually crippled and lame. He does not*

bother about the ones that are spiritually dead; he already has those.

A former friend wrote:

Oh-h-h Irene, you poor girl. Two years ago, you were leading a completely normal, carefree, girlish life, surrounded by family and friends—with peace, blessed peace, reigning in your heart and life.

I am certain you never did imagine what would happen in 1996. Oh that Satan: Why does he have to go around gleefully wrecking people's lives? The tears threaten to overwhelm me.

She was wrong on several accounts. Mine was not a normal, carefree life among the Amish. I clearly didn't have peace in my heart. And I will never view my departure as the handiwork of Satan.

But she unwittingly was right about one thing. I still longed to have closure with my mother and father, and in that way I was still seeking a peace. My first visit home when I saw my brothers left me feeling empty—as if I had not accomplished what I'd set out to do. I had faced my detractors squarely, but I had not confronted their leader. My father. And I had not been able to comfort my mother.

In the two and a half years since my first trip to Kalona, I had thought often about returning. And every time I did so, I shuddered at the torture that awaited me. I had learned in corresponding with my father that there was no room for debate when it came to what I had done. He considered me a disobedient child, and he would be full of preaching, disapproving looks, and tears.

The first visit was such an effort, such a mental drain, it took the better part of those two and a half years to build myself back up again. To repair my mettle for battle.

I also remained distraught that my family had not visited me in all that time. They had offered—if I gave them permission. But the truth is, I had invited them repeatedly. If they wouldn't come, why should I go?

Further, my family had made it difficult for me to visit on special occasions. In 1998, they had a mid-week Christmas dinner for my mother's family—in August. I wasn't told about it until it was too late for me to make travel arrangements. The following year, I wasn't invited to my brother Aaron's wedding.

Many times, Ottie would say to me, "If you want to go see your mother and father, I'll take you."

More often than not, I would reply, "I don't want to talk about it."

In the back of my mind, I suspected the time would come when I'd have to go back. But I kept putting it off. I kept working on strengthening myself and my marriage, and on learning everything there was to know about my new environment, which continued to offer silly oddities as well as uplifting discoveries.

One of the strangest things struck me during Horse Cave Days, a little come-as-you-are downtown festival with live bluegrass music, sidewalk sales, a two-dollar-pet-a-boa cage, lemonade, and a NASCAR race car. But what caught my attention was a makeshift dunk station, where a beefy teen sat on a folding chair beneath a bucket of water and let people douse him with cold water by hitting a bull's-eye with a softball. Why, I wondered, would anyone want to do that? What was the purpose? Where was the thrill? I was completely befuddled.

Conversely, one of my greatest finds was the spectacle of the English Christmas.

The Old Order Amish are very reserved about celebrating Christ's birthday, and at the farmhouse we unceremoniously got one gift apiece on Christmas morning. Unwrapped.

The gifts were always things we could use, not necessarily things we wanted. Shoes, boots, gloves, a coat. And except for the shoes, they were all homemade.

Sometimes we'd also have candy or orange slices as treats. But there were no Christmas trees, no Santa Claus, no decorations.

People in the young folks group were a little more demonstrative, and they would exchange wrapped presents and go caroling throughout the community.

The difference had everything to do with age. Once an Amish person becomes an adult, gets married, and has family, they are expected to be more serious and less frivolous. Christmas, at that point, takes on a somber, workmanlike tone. The weight of the world is upon a person.

The greed with which some English children accord Christmas was unsettling at first. Tearing the paper off presents one after another, never bothering to savor the joy of each gift and never bothering to preserve the colorful wrappings. Associating love with the expense of an item. Being enamored with useless novelties rather than presents with purpose.

But there is so much more that is wonderful about the holiday, and the preparations alone were cause for glee. The wrapping of presents. Putting up the Christmas lights around the house. Decorating the tree. Making fudge, cookies, pies, and bread. All of it done to the sound of Christmas music wafting from the stereo.

Then there are the holiday movies and TV shows, and the downtowns bedecked with banners and lights and window displays.

The Amish believe the English have commercialized Christmas, and that they worship Santa Claus, not Jesus. But this commercialization, if that's what it is, is a glorious way to herald the coming birthday of Christ. Further, English churches don't preach about Santa Claus, they focus on the Savior. Santa Claus is nothing more than a warm, harmless, comforting tale for children.

My first Christmas in Kentucky, I decorated the house from one end to the other with lights, tinsel, and nativity scenes.

I wasn't the easiest person to buy presents for. I was, after all, accustomed to getting only one and it was always a practical gift. Ottie would ask me what I wanted for Christmas and I would reply: "I don't know what I want. Just get something."

I was also appalled when we'd buy four or five gifts for each family member.

"We're going to spend two thousand dollars on Christmas, and they've already got everything they need," I complained.

But for the most part, I resisted my urges and immersed myself in the spirit of the season. And on Christmas Day, I gave Ottie a watch and he gave me a gorgeous diamond heart necklace. They were my first diamonds and, like many English women, I spent a considerable amount of time watching them sparkle in the sun.

I even learned, eventually, that not all things must be purposeful. One day when walking through a store, a giant plastic sunflower I had passed began singing "You Are My Sunshine."

I persuaded Ottie to buy it and propped it in our living room, where it sang to everyone who walked by. Sometimes, I even sang along with it.

Those were the good days, and there had been many others like it since leaving Kalona. Marrying Ottie and being befriended by his family. Growing closer to Christ. Getting to know Rev. Bettermann. Having the ban lifted. Learning to drive. Acquiring my GED. Learning to laugh more often. And making many new friends.

For all of it, though, the distance between me and my family had remained a painful, almost constant thorn.

Growing up Amish, a person's chief concerns—beyond an allegiance to God—are complying with the rules, being approved of by the people around you, and blending in. In short, being accepted.

And despite my acclimation to the outside world, despite my acceptance in my new home, I still felt the tug of wanting to belong to my birth family—in some small way, at least.

The sores on my mother's legs were the impetus that provided an urgency for a visit, although it was a kind of compassion I apparently shared alone.

In the spring of 2000, I went to the hospital to have a benign lump the size of a golf ball removed from the left side of my abdomen. I had written my family telling them of the impending operation, but had heard nothing in return.

Inasmuch as it was my first surgery and there was always the chance the growth might be malignant, it would have been nice to have received some encouraging words from my family. Something to tell me they were thinking of me, praying for me.

My mother sent a get-well card a week after the surgery. But as it turned out, my biggest support system was Ottie's family, members of the Lutheran congregation, neighbors, and God.

Rev. Bettermann visited Ottie and me in preop before the surgery, held our hands, and prayed, letting us know God was with me.

Faye stayed at the house my first night out of the hospital, did laundry for the next couple of days, washed dishes, and ran errands, and neighbors Don and Betty Gumm brought us food. People from the church, meanwhile, called with offers to clean our house.

God, as always, was there, too. Guiding me. Enlightening me. Comforting me.

There is a passage from Isaiah (41:10) that has always been by my side: "Fear thou not; for I am with thee: be not dismayed; for I am thy God; I will strengthen thee; yea, I will help thee; yea, I will uphold thee with the right hand of my righteousness."

It was armed with this promise and Ottie's unwavering devotion that I set out for Kalona again, this time with my aunt and uncle, determined to set the record straight.

Ottie, alas, stayed behind. To demonstrate to my parents that I was free to come and go as I pleased. To show them there was no basis for the fanciful rumors the Amish had drummed up.

I would have loved to have had him with me. We had, after all, become joined at the hip and heart. And, we had never been apart for more than a day since leaving Kalona.

Twenty-Five

February 5, 2000

It took awhile, but eventually the conversation with my mother mercifully turned from the wild rumors to her health.

The subject of rumors came up only twice again. In town, we heard that Ottie was in prison for child molesting, which of course wasn't true. Later in the day, after my father had returned to the farm, he also mentioned a few tall tales.

For the moment, though, it was nice talking to my mother alone, unencumbered by my father's stifling presence. My fourteen-year-old brother, Earl, would occasionally stand nearby and listen. But he said no more than six words, bashful as he is. It didn't help, I suppose, that I mentioned his voice had dropped since I'd last seen him.

My mother threw a few guilt trips my way.

"Oh, it's just such a shame," she said at one point. "Such a waste."

But much of our conversation centered on her guilt about my leaving.

"If we could have gotten to you sooner after you left, would you have changed your mind?" she asked.

"No," I said. "I had my mind made up."

"What if I had never told you about my life?" she asked, alluding to things she'd told me about her life before she'd gotten married.

"No," I said. "That wouldn't have changed anything, either. I've seen enough on my own in this house without you telling me."

Less than an hour after I arrived, a driver took Mom and Earl to Iowa City to get eyeglasses for my brother. I agreed to return later in the day, along with my aunt and uncle. Then we went to see my married brothers, Wilbur, Elson, and Aaron.

The visit with Elson went well enough, the visits with Wilbur and Aaron less so.

While Aaron and his wife, Martha, showed me around their new home, they began talking to me about returning to the Amish. Once we'd reached their bedroom, they offered a surprise.

"You can go get it now if you want," Aaron told his wife.

"Okay," she said, and she walked to the dresser, retrieved a letter and card I had sent to Aaron for his birthday, and handed them to her husband.

In a rare show of affection, Aaron put his arm around me and said, "You know how we feel about this and I guess you know why we can't accept it."

It was as if he felt bound to do something he didn't want to do, and I privately felt sorry for him.

"That's up to you," I told him. "It's your choice."

For the rest of the house tour, we sparred about Amish beliefs and traditions, going over some of the same issues I had addressed with my father and would later discuss with my former bishop, Elmer T. Miller, the uncle who had formally notified me of the ban.

Aaron and Martha brought up the matter of English churches letting their parishioners go to war.

"You can look at it that way, but if no one in this country had ever fought, there would be no Amish, no Christians," I said. "Hitler would have seen to that a long time ago.

"Besides," I continued, "did you ever think that the Amish came to this country seeking religious freedom, and that you can enjoy religious freedom because of those who gave their lives?"

"Well," Aaron said, "we'd like to think that prayer had a lot to do with stopping that war."

"I'm sure it did," I said. "But what about all the mothers' sons who died?"

He didn't answer. Instead, he posed another question:

"What's so bad about the Amish, anyway?"

I wanted to say: "What's so good about them?" But I didn't.

"There's good and there's bad," I said.

More of the latter came when we returned to the farm. Seconds after I entered, looking every bit liberal Amish with a jumper, blouse, and small head covering, my father arrived and took a long, stunned look at me. I tried to reach out and hug him, much the way I had done with Ottie's father years before, but he pushed me away, swept across the living room to my mother, and began weeping.

"Oh Irene!" he cried. "I just think it can't be. We wanted to see you, but not like this. Not like this."

When his crying eventually subsided, the room turned to deafening silence, each of us awkwardly trying to make conversation—and failing miserably.

Finally, my father announced that he would summon my uncle to talk to me. Or better put: He asked Earl and Benedict to summon him.

"I didn't come to see the bishop," I said.

"What's wrong?" my father asked. "You don't want to talk to him?"

The truth was, I didn't want to talk to him, was even slightly afraid of talking to him. But I didn't dare let my father know.

"I don't care," I told my father. "I can talk to him, though I already know what he's going to say. But I'll tell you this: If he brings other people with him—a bunch of ministers, for instance—I'm not staying. I'm just letting you know that right now."

My father didn't exactly agree to the terms, but he didn't exactly disagree either. He just kept asking the boys to fetch my uncle.

"Well, you gonna go tell him she's here?"

The boys stood motionless, without uttering a word. They seemed hypnotized by the moment—or, more likely, frozen with fear.

"Why don't you go tell him?" my father urged again.

And again, there was no response.

The third time, my father's urging became a command.

"Take the buggy," he told Earl.

And off he went.

The bishop, a short, lean man with sad eyes, talked with me on a side porch for forty-five minutes.

"Well, you've heard you're in the ban," he began.

"Am I?" I asked. "The Lutheran church lifted the ban. The minister of the church wrote you a letter telling you I was no longer in the ban, but you never answered it."

"Oh?" he said with mock astonishment. "I didn't think that's what the letter said. Still, you know how we feel about you marrying a divorced man."

"Yeah, I do. But which is the greater sin: Living together and fighting and hating each other, or getting a divorce?"

"Well, I guess you're talking about your parents now, but. . . ."

He didn't, or couldn't, finish, and our talk disintegrated into trading Bible verses to support our arguments. It was an unpleasant way to conduct a debate.

"You know," I said toward the end, "we could stand here and do this all day and not get anywhere."

"We could," he agreed.

And that was pretty much that. An impasse.

Rev. Bettermann wrote another letter to my uncle, but it wasn't answered, either. My family, meanwhile, with the exception of Elson, slowed the flow of letters to me after my coauthor told them he had been reading them.

I still go to the mailbox every day, hoping, praying that I will hear something from Kalona. I don't understand how it is that a family can turn its back on a daughter whose only transgression was building a more solid relationship with God, whose only pursuit continues to be the truth, and who's need to learn, thankfully, was greater than any fear.

I hold on to my faith that God will continue to lead me in the right direction and that he will one day help my family understand, if not accept, my decision. Although my visit didn't have the outcome I had hoped for, I know that in their own sometimes small ways, despite being tied to centuries of tradition, members of my family do love me even if they can't come right out and say it.

There is evidence in a few precious moments tender and fleeting.

When I was about three, my father held me in his arms and rocked me after I'd burned my hand on a stove.

In October 2000, several relatives on my mother's side, some of whom were still Amish, briefly—and unexpectedly—visited our house in Horse Cave.

My mother and father sent me a birthday card early in 2001 that said "Somebody Misses You," accompanied by conciliatory letters that skipped the fire and brimstone.

And when I left the Kalona farm after my February visit, I hugged my mother and whispered in her ear, "I love you."

My mother whispered back in Pennsylvania Dutch, the language void of the word "love."

"Ich du dich, aw," she said.

I do you, too.

Special Thanks

Rick Farrant

I owe a great debt of gratitude to those people who encouraged my talents, confirmed my compassion and, in some cases, literally saved my life.

Among them:

Susie, Amber, Tyler, and Nicholas Farrant, who give me love, inspiration and unwavering support, no matter my transgressions. I love you all.

Ottie and Irene Garrett, who trusted me implicitly and have became good friends during the course of our work together.

Phyllis Rogers and Joan Warner-Engel, who believe more in me than I do in myself.

Sandy Thorn Clark, Steve Penhollow, Jason Stein, and Carroll Ann Moore, who read this manuscript and provided invaluable feedback. You are journalists, writers, and colleagues of the highest order.

Jim Plessinger, Gay Cook, Clint Wilkinson, Mike Whitehead, Chuck Morris, and Lawrence Pettit, who saw a fire in me and took the time to help fan it.

Jim Carrier, whose words "to someone who also dreams of words beyond newspapers" have stuck with me all these years, and to countless others whose written words have buoyed me.

Gideon Weil and the fine folks at Harper San Francisco, who believed Irene's story was worth telling.

Bill Gaither, Rebecca St. James, Ginny Owens, and Father Joseph Girzone, who have been brief but powerful spiritual spark plugs.

All of my other interview subjects along the way—from John Mellencamp to Debbie Reynolds—who shared their deepest thoughts and sterling wisdoms.

James Perkins, an estranged but once-important friend who brought light to an unhappy childhood, and Ruby McGown and Rhea Edmonds, newfound friends who prosper because they believe and dream.

The Robys, Wilsons, and Gilchrists, who made an adopted child feel like a full-fledged member of the family.

Dr. Basil Genetos, who repaired a broken heart and kept it running long enough to see this project through. Keep it up, doc.

And, finally, God, who led me to his altar in my darkest hour and taught me how to receive and revere him.

Special Thanks

Ruth Irene Garrett

I hope this small gesture shows my gratitude to so many people who have given me support, kindness, understanding, and love.

Our friends and family in Christ, Rev. James A Bettermann and family, The Holy Trinity Lutheran Church, Mark and Kathy Tooley and daughters, Ron and Kathy Zielke, Martha Zielke, Arthur and Fletta Diamond, Neva and the late Bob Gielow, Caroll and Susan Smiley, Ann Padilla, Starlot and James Pierce, Ray and Carol Penhollow, Dallas Flowers, Mark and Marion Easterday and family, Connie and Bill Beach, "Mama" Ellen, Patrick and Christina Stewart and family, Robert and Judy Stokes and daughters, Nanette Varian of *Glamour* magazine, and Beth Polson of Polson Productions.

My Aunt Edna and Jake Schmucker and those in my family who had the courage and kindness to come visit our home.

Betty and Don Gumm, Mike, and the late Freda Jewell for their friendship and letting us live on their "Home Place."

My husband's family and his children, I love you all. A special thank you to my father-in law, Ottie Garrett Sr. Our home is beautiful.

A special thank you to Monte Garrett and his four friends, Fred McDonald, Chris Chase, Richard Hoover, and the late

Wayne Wilson for their midnight escapade, moving Ottie's household belongings from Iowa to Kentucky in record time.

Gideon Weil, Liz Perle, and Harper San Francisco for publishing my story.

In closing, my family and my Amish heritage will remain a part of me and I love them all unconditionally.